Building Floors, Walls and Stairs

Fine Homebuilding®

Building Floors, Walls and Stairs

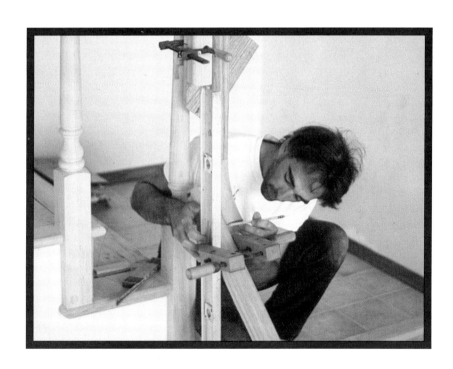

The Taunton Press

Cover photo by Sebastian Eggert

©1988 by The Taunton Press, Inc.

First printing: December 1988
International Standard Book Number: 0-942391-12-8
Library of Congress Catalog Card Number: 88-50567
Printed in the United States of America

A FINE HOMEBUILDING Book

FINE HOMEBUILDING is a trademark of The Taunton Press, Inc.,
registered in the U.S. Patent and Trademark Office.

The Taunton Press, Inc.
63 South Main Street
Box 355
Newtown, Connecticut 06470

CONTENTS

INTRODUCTION

No aspect of building requires more skill and produces more pride than finish carpentry. It creates all those parts of a house that can be seen and touched, work that demands precision and tireless attention to detail. Whether you're spackling joints in drywall, sanding an oak floor or installing a newel post and balustrade, you know that every flaw will flash forth in the finished work. The trick is to keep flaws to a minimum, and to know how to fix them when they happen.

This book contains 39 articles from the back issues of FINE HOMEBUILDING magazine written by and about professional builders.* They describe how they work, and explain how they've managed to achieve excellent results by practical means. Fifteen of the articles deal with designing and building stairways, while the others cover a myriad of subjects from wainscoting to plaster-casting rosettes.

—John Lively, editor

*The six volumes in the Builder's Library are from FINE HOMEBUILDING magazine numbers 1 through 46, 1981 through early 1988. A footnote with each article tells when it was originally published. Product availability, suppliers' addresses, and prices may have changed since then.

The other five titles in the Builder's Library are *Tools for Building; Frame Carpentry; Building with Concrete, Brick and Stone; Building Doors, Windows and Skylights; Building Baths and Kitchens.*

Designing and Building Stairs

Stairways can be minimal or very elaborate, but they're all based on simple geometry and accurate finish work

by Bob Syvanen

Today, few people can afford the time and expense of building classical 19th-century stairways. Tastes have changed too, and as a result, stairs have become simpler. But the principles of stair design and construction are the same as they've always been, and so are the skills that the builder must bring to the task. If modern stairs aren't ornate, they still remain a focal point of a house—your work is out there for all to see.

One of the problems builders face in cutting stairs is getting rusty. Unlike the master stair-builders of the past, on-site carpenters typically build only a few stairways a year. Fortunately, designing a stair and laying out the stringers uses the same language and concepts as roof framing. Fitting treads and risers, on the other hand, is nothing more than simple, if demanding, finish work.

Basic stair types—The simplest stairway is a cleated stairway, which relies on wood or metal cleats fastened to the carriages to support the treads. You could use a cleated stairway for a back porch or cellar, but count on repairing the ever-loosening cleats. Wood cleats are typically 1x4s, and screwing them in is a big improvement over nails. But angle-iron cleats will last longer.

Another open-tread stairway—one without risers—uses dadoed carriages. The treads have ½ in. or more bearing on the inside face of the carriages, and they are either nailed or screwed in place. I use a circular saw, and set the depth of cut to half the thickness of the carriage, to make the parallel cuts. Then I clean up the bottom of the dado with a chisel. A router and a simple fixture built to the stair pitch and clamped to the carriage will also do a quick, neat job.

In the past, dadoed carriages were used mostly for utility stairs for porches, decks and the like, but more and more I'm asked to build open-tread oak stairs that have to be nearly furniture quality. One type, the stepladder stair, can be used when limited space doesn't allow any other solution (for more, see p. 14).

On finished stairs, the dado is usually stopped (that is, its length is limited to the

Stairbuilding in its highest forms requires the conceptual skills of a roof framer, and the fitting talents of a cabinetmaker. But simple open tread runs like the basement stair at left only require understanding the basics.

From *Fine Homebuilding* magazine (October 1983) 17:56-62

width of the tread), and squared up at the end with a chisel. If you begin the dado on the front of the carriage, leave the tread nosing protruding slightly. If you start the dado at the back of the carriage, the treads will be a little inset. They can look nice either way.

Other stairways use cut-out carriages. The simplest of these is a typical basement stair where the rough carriages are exposed, and risers are optional. The most complicated is a housed-stringer stairway. It is based primarily on patient and accurate work with a router. The stringer is mortised out along the outline of the treads and risers with a graduated allowance behind the riser and tread locations for driving in wedges. Adjustable commercial fixtures or wooden shop-made templates are used to guide the router. Although the first stairway that I built by myself on my own had a housed stringer, I won't try to give complete instructions here.

A finished stairway that still requires patient finish work, but is much less tedious to build, uses cut-out carriages hidden below the treads, and stringboards or skirtboards as the finish against the stairwall. The treads and risers are scribed to the skirts. The example I'll be using to describe final assembly is one of these that also has an open side, which uses a mitered stringer. This side of the stair requires a balustrade—handrail, balusters, and newel posts—but that's a separate topic.

Designing a stairway—No matter what style stairway you want to build, the design factors that you'll need to consider are comfort, code, safety and cost. Comfort gives the ideal conditions for good walking, and code dictates what you can and cannot do. Safety is largely a matter of common sense, and cost limits your grand ideas.

Although comfort is very much a subjective notion, there are three objective factors to consider in designing a stair—stair width, headroom and the relationship between the height of the riser and the width of the tread. Of the three, stair width is the easiest to deal with. Most building codes require utility stairs to be at least 2 ft. 6 in. wide and house stairs to be 3 ft. from wall finish to wall finish, but 3 ft. 6 in. feels a lot less restrictive. However, if the stairway gets beyond 44 in. in width, most building codes require a handrail on each stair wall.

Headroom is not quite as simple, although most codes agree that basement stairs need a minimum of 6 ft. 6 in., and house stairs need at least 6 ft. 8 in. This measurement is made from the nosing line (for a definition, see *headroom* in the glossary on the next page) to the lowest point on the ceiling or beam above. A lack of headroom is most noticeable when you're going down the stairs, because you are walking erect and bouncing off the balls of your feet. Ideal headroom allows you to swing an arm overhead going downstairs, but this requires a clearance of nearly 7 ft. 4 in. Headroom can be increased by enlarging the size of the stairwell, decreasing the riser height, or increasing the width of the treads (tread

width is the distance from the front to the rear of the tread).

Certain combinations of riser and tread are more comfortable than others. Most codes set limits—a maximum rise of 8¼ in. and a minimum tread width of 9 in.—but these are based on safety, not comfort. A 7-in. rise is just about ideal, but it has to be coupled with the right tread width.

The timeworn formula for getting the tread width and riser height in the right relationship is: riser + tread = 17½ (a 7-in. riser + a 10½-in. tread = 17½ in.) Another rule of thumb is: riser × tread = 75 (7 × 10½ = 73.5, which is close enough). Still another formula is: two risers + one tread = 24 in. I've always found the first formula the easiest. All of them establish an incline between 33° and 37°. This creates a stairway that is comfortable for most people.

More considerations—Each tread should project over the riser below it. This projection, or nosing, should be no more than 1¼ in. and no less than 1 in. In open-tread stairways, a tread shouldn't overlap the tread beneath it by more than ½ in. The nosing adds to the area where your foot falls, but doesn't affect the rise-run dimensions. The top of a handrail should be between 30 in. and 34 in. above this nosing, with a 1½-in. clearance between the handrail and the wall. If the top of the stairway has a door, use a landing at least as long as the door is wide.

Keep in mind that people aren't the only things moved up and down stairs. I once lived in an old Cape Cod house that had a stairway with 10½-in. risers, 6½-in. treads, and not too much headroom. You could negotiate it if you exercised a little caution, but moving heavy furniture up and down was another story.

The size of a stairwell is based on the riser and tread dimensions, and on how much headroom you need. A typical basement stair can be gotten into a rough opening 9 ft. 6 in. long by 32 in. wide. A main stairwell should be a minimum of 10 ft. by 3 ft.

Framing a stairwell isn't complicated and is usually defined in the local building code. If the long dimension is parallel to the joists, the trimmer joists on each side are doubled, as are the headers. Similar framing is required if the long dimension is perpendicular to the joists (drawings, above right). If the header isn't carried by a partition below, it will have to be designed for the load.

Making the calculations—Careless measuring can get you in a lot of trouble when you're building stairs. Although the initial figuring may seem a little theoretical, you'll soon be doing some fussy finish work based on these calculations and the resulting carriages. First, check both the stair opening and the floor below for level. If either is out of level, determine how much. You'll have to compensate for it later. This problem occurs most often with basement slabs.

There are many ways to lay out stairs; the following system teamed with a pocket calcu-

Framing stair openings

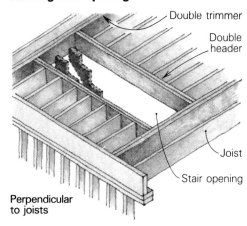

Perpendicular to joists

Double trimmer
Double header
Joist
Stair opening

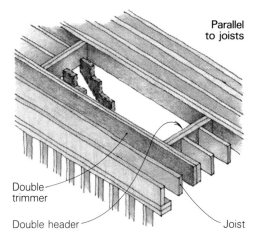

Parallel to joists

Double trimmer
Double header
Joist

lator works well for me, even though I have to scratch my head to recall what I did last time. Start with the measurement that has already been determined, the finished floor-to-floor height. I'll use 108 in. in this case. Then, just for a starting point to get you close, divide by 14, the average number of risers used in residential stairs. This gives a riser of 7.71 in., which is a little high. Adding another riser will reduce this measurement some: 108 in. ÷ 15 = 7.20 in. Sounds good. You now know how many risers you'll be using and how high they are.

To get the width of the treads (remember, this means the front-to-back measurement of each step), use the rise-plus-tread formula in reverse: 17.5 in. − 7.2 in. = 10.3 in.

All stairs have one more riser than treads. This is because the floors above and below act as initial treads, but aren't a part of the stair carriage calculations. You'll have to keep reminding yourself of this when you lay out the carriages. I don't know a good way of remembering this, and I have resorted to drawing a sketch of a couple of steps and using it to count the difference. In the example we're using, then, there are 14 treads and 15 risers.

The last calculation is the total run—the length of the stairway from the face of the first riser to the face of the last riser. This is simply the total number of treads multiplied by their width: 14 × 10.3 in., or 144.2 in. With this figure you can check to see whether the stairway will fit in the space available. Although I now use a calculator for the math, my main tool used to be a stair table like the

A Glossary of Stair Terms

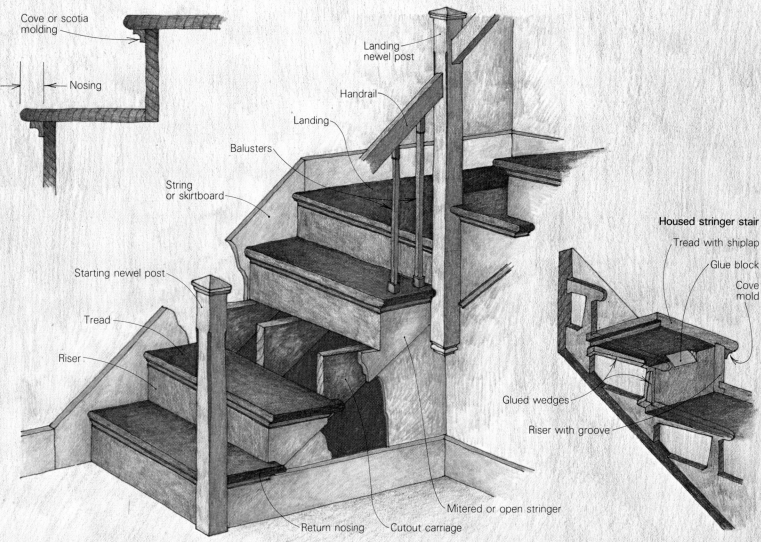

Cove or scotia molding

Nosing

Landing newel post

Handrail

Landing

Balusters

String or skirtboard

Starting newel post

Tread

Riser

Housed stringer stair

Tread with shiplap

Glue block

Cove mold

Glued wedges

Riser with groove

Mitered or open stringer

Return nosing

Cutout carriage

Balusters—The posts or other vertical members that hold up the handrail, usually two per tread.

Balustrade—The complete railing, including newel posts, balusters and a handrail. Most of these parts are available as stock finished items at lumberyards.

Carriages—Also called *stair stringers, stair horses* or *stair jacks*. They are the diagonal members that support the treads. Carriages can either be *finish stringers* or rough stringers—for an outside stairway, or for an inside stairway hidden from view. *Rough carriages*, whether they are *cut-out carriages* or just *dadoed* or *cleated stairs*, are made of 2x10 or 2x12 softwood lumber. Finish stringers are usually made of ¾-in. or 1⅛-in. stock. They either can be cut out (an *open or mitered stringer*) or routed (a *housed stringer*).

Closed stairway—Stairs with walls on both sides. In this case a *wall stringer*, whether it is a *housed stringer* or just a *stringboard*, is nailed to each wall. Closed stairways use handrails, not a balustrade.

Finished stairway—Any of several interior stair types that have risers, treads, stringers and a handrail or balustrade.

Handrail—This rail runs parallel to the pitch of the stairs. It's held by balusters or brackets.

Headroom—The vertical distance from the lowest point of the ceiling or soffit directly above the stair to the *nosing line*, an imaginary diagonal connecting the top outside corners of treads. Most codes require at least 6 ft. 8 in. for stairs in living areas, and 6 ft. 6 in. for basement utility stairs.

Housed stringer—The profile of the treads, nosing and risers is routed into a finish stringer. Extra room is left for wedges to be driven and glued in between the stringer and the treads and risers. Rabbeted and grooved risers and treads are also used.

Landing—A platform separating two sets of stairs.

Newel post—The large post at the end of the handrail. There is a *starting newel* at the base of the stairs, and a *landing newel* at turns.

Nosing—The rounded front of the tread that projects beyond the face of the riser 1 in. to 1¼ in. In the case of *open-tread stairways*, it shouldn't exceed ½ in. In most cases, the nosing is milled on the tread stock. On open stairways, a half-round molding called *return nosing* is nailed to the end of the tread.

Open or mitered stringer—This is a cut-out finish stringer used in open stairways. The treads carry over the stringer, but the vertical cut-outs on the carriage are mitered with the risers at 45°.

Open stairway—This can be open on one or both sides, requiring a balustrade. In finished stairways, the open sides will use a *mitered* or *open stringer*.

Rise—The height of each step from the surface of one tread to the next. Just as in roof framing, this measurement is sometimes called the *unit rise*. Many codes call for a maximum rise of 8¼ in. The height of the entire stair, from finished floor to finished floor, is the *total rise*.

Riser—Describes the rise of one step. It is also a stair part—the vertical board of each step that is fastened to the carriages. Risers for a *housed stringer stair* are rabbeted at the top to fit the tread above, and grooved near the bottom for the tread below. Other stairs use 1x square-edged stock. *Open-tread stairs* don't have risers.

Run—Also called *unit run*, this is the horizontal distance traveled by a single tread. A 9-in. run is the code minimum for main stairs. *Total run* is the measured distance from the beginning of the first tread to the end of the last tread—the horizontal length of the entire stairway.

Stairwell—The framed opening in the floor that incorporates the stairs. Its long dimension affects how much headroom the stair has.

Stringboard—Diagonal trim, not used to support the treads, that is nailed to the stair walls. Finished treads and risers butt these. Often called *skirtboards, backing stringers,* or *plain stringers.*

Tread—It is both the horizontal distance from the face of one riser to the next, and the board nailed to the carriages that takes the weight of your foot. Exterior stairs typically use 2x softwood treads. Interior stairs use either 1⅛-in. hardwood stock milled with a rabbet and groove to join it to the risers, or 1³⁄₁₆-in. square-edged stock. Both are usually nosed.

Winder—Wedge-shaped treads used in place of a landing when space is cramped, and a turn is required in the stairway. Many building codes state that treads should be at least the full width of the non-winder treads, 12 in. in from their narrow end; or that the narrow end be no less than 6 in. wide.

—*Paul Spring*

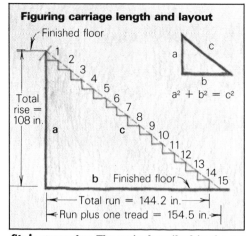

Figuring carriage length and layout

Finished floor

Total rise = 108 in.

$a^2 + b^2 = c^2$

Total run = 144.2 in.

Run plus one tread = 154.5 in.

Total rise*	Number of risers	Riser height	Tread width	Total run
8'0"	12	8"	9"	8'3"
	13	7⅜" +	9½"	9'6"
	13	7⅜" +	10"	10'0"
8'6"	13	7⅞" −	9"	9'0"
	14	7⁵⁄₁₆" −	9½"	10'3½"
	14	7⁵⁄₁₆" −	10"	10'10"
9'0"	14	7¹¹⁄₁₆" −	9"	9'0"
	15	7³⁄₁₆" +	9½"	11'1"
	15	7³⁄₁₆" +	10"	11'8"
9'6"	15	7⅝" −	9"	10'6"
	16	7⅛"	9½"	11'10½"
	16	7⅛"	10"	12'6"

*height from finished floor to finished floor

Stair geometry. The stair described in the text is shown above left. Plugging total run and total rise into the Pythagorean theorem gives the required carriage length. Based on an ideal riser and tread, the stair chart, above right, gives the number of risers and total run for a given total rise.

one above. I still use one for quick reference in the planning stage. If the run is too long, make the treads narrower, or eliminate a riser and a tread. Either way, you'll need to run a new set of calculations.

Layout—To lay out the carriages, you first need to know what length stock to buy. If you've got a calculator, it's easiest to use the Pythagorean theorem $(a^2 + b^2 = c^2)$. But you must add the width of an extra tread to your total run to get enough length for the bottom riser cut (drawing, above). In this case, a is the total rise of 108 in., and b is the total run of 144.2 in. plus a 10.3-in. tread. The hypotenuse, c, is 188.5 in. So you'll have to buy 16-footers to allow for cutting off end checks, and avoiding large knots with the layout.

Most cut-out carriages are 2x12s because you need at least 3½ in. of wood remaining below the cut-outs for strength; 4 in. is even better. Douglas fir is the best lumber for the job because of its strength. You will need a third, or center, carriage if the stair is wider than 3 ft. with 1½-in. thick treads, or wider than 2 ft. 6 in. with 1⅛-in. thick treads.

Once you have marked the edge of one of the carriages with the 188.5-in. measurement, you are ready to lay out. I step off equal spaces with dividers (photo top right) before marking the riser and tread lines with a square. Some carpenters simply step off the cut-out lines with a square, but I don't like the accumulated error you can get this way. A deviation of more than ¼ in. between the height of risers or the width of treads can be felt when walking a flight of stairs. For this reason, it's also a violation of code.

The dividers I use are extra long. You can improvise a pair by joining two sticks with a finish nail for a pivot, and a C-clamp to hold them tightly once they're set. The easiest way to find the spacing is to locate 7.2 in. (the riser height) and 10.3 in. (the tread width) on a framing square, and set the divider points to span the hypotenuse, which is 12.56 in. (a strong 12½ in.). No matter how careful you are in setting the dividers, it will take a few trial-and-error runs before you come out to 15 even spaces. Once you do, mark the points on

the edge of the carriage. These represent the top outside corner of each tread, less the profile of the nosing.

To draw the cut-out lines, use either a framing square or a pitchboard. Most carpenters use the square, but a pitchboard can't get out of adjustment. You can make one by cutting a right triangle from a plywood scrap. One side should be cut to the height of the riser, and the adjacent side to the tread width. A 1x4 guideboard should be nailed to the hypotenuse. Align it with the marks on the carriage and use it to scribe against.

If you are using a square, set it on the carriage so that the 90° intersection of the tongue and body point to the middle of the board. Along the top edge of the carriage, one leg should read the riser increment and the other, the tread increment. Use either stair-gauge fixtures (stair buttons) or a 2x4 and C-clamps to maintain the correct settings when the square is in position against the edge of the carriage. Then, holding the square precisely on the divider marks, scribe the cutlines for each 90° tread-and-riser combination (photo second from top).

Dropping the carriages—One of the most difficult things about stairs is adjusting the carriages for the different thickness of floor finish, which can throw off the height of the bottom and top risers. Any difference should be subtracted from the layout after the tread and riser lines are marked, and carefully double-checked before you do any cutting. What you marked on the carriage is the top of the treads, but since you will be nailing treads to the carriages, you need to lower the entire member enough to make up for the difference. This is called *dropping the carriages*. If they sit on a finished floor, such as a concrete basement slab, the bottom riser will need to be cut shorter by the thickness of a tread (drawing A, next page, top right). This will lower the carriage so that when the treads and upper floor finish are added, each step will be the same height. The bottom riser will have to be ripped to a narrower width. If the treads and floor finishes are of equal thickness (B), and the carriage sits on the subfloor at the bottom,

Laying out and cutting the carriages. Starting with the top photo, Syvanen uses large dividers to mark the intersections of tread and riser lines on the front edge of the carriage. From these marks, he scribes the cut-out lines using a framing square fitted with stair-gauge fixtures. With a circular saw, he begins cutting out the carriages by notching the bottom riser for the kickplate that will anchor it to the floor. Syvanen uses a handsaw to finish the cutting. Cut-out carriages are usually made of 2x10s or 2x12s, since there should be at least 3½ in. of stock between the bottom edge of the carriage and the cutout.

Photos: Bob Syvanen; Illustrations: Frances Ashforth

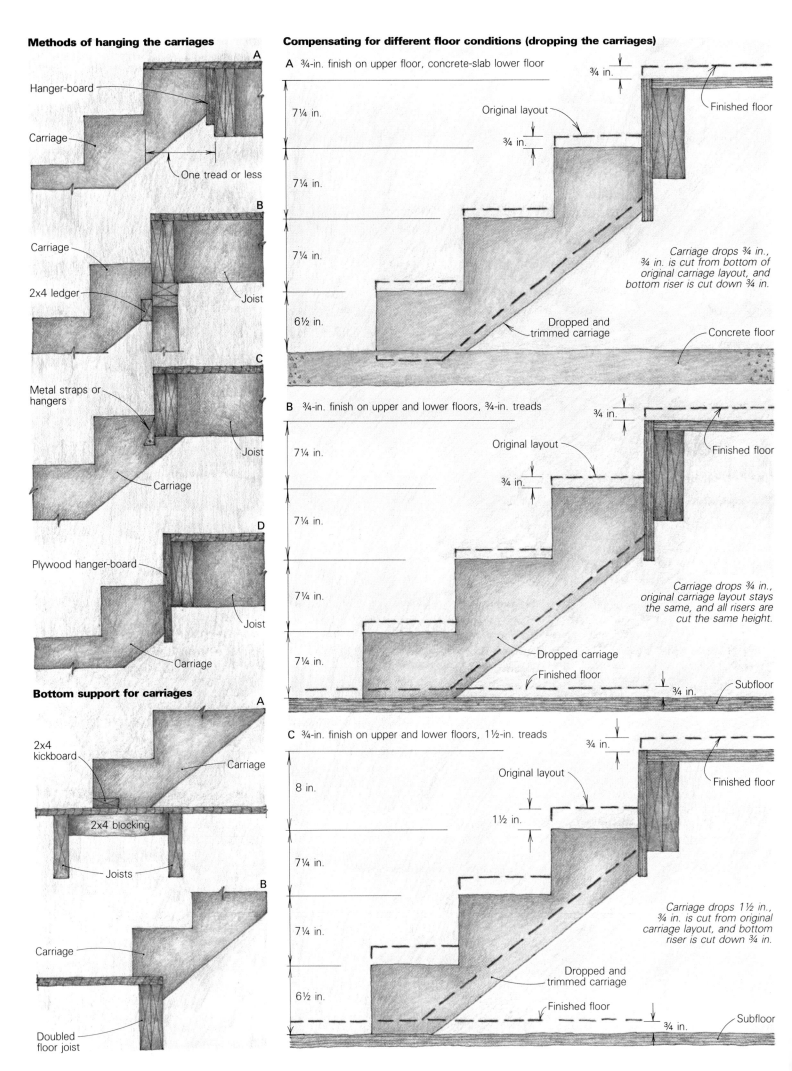

Methods of hanging the carriages

A
Hanger-board

Carriage

One tread or less

B
Carriage

2x4 ledger

Joist

C
Metal straps or hangers

Joist

Carriage

D
Plywood hanger-board

Joist

Carriage

Bottom support for carriages

A
2x4 kickboard

Carriage

2x4 blocking

Joists

B
Carriage

Doubled floor joist

Compensating for different floor conditions (dropping the carriages)

A ¾-in. finish on upper floor, concrete-slab lower floor

¾ in.

Finished floor

Original layout

¾ in.

7¼ in.

7¼ in.

7¼ in.

6½ in.

Carriage drops ¾ in., ¾ in. is cut from bottom of original carriage layout, and bottom riser is cut down ¾ in.

Dropped and trimmed carriage

Concrete floor

B ¾-in. finish on upper and lower floors, ¾-in. treads

¾ in.

Finished floor

Original layout

¾ in.

7¼ in.

7¼ in.

7¼ in.

7¼ in.

Carriage drops ¾ in., original carriage layout stays the same, and all risers are cut the same height.

Dropped carriage

Finished floor

¾ in.

Subfloor

C ¾-in. finish on upper and lower floors, 1½-in. treads

¾ in.

Finished floor

Original layout

1½ in.

8 in.

7¼ in.

6½ in.

Carriage drops 1½ in., ¾ in. is cut from original carriage layout, and bottom riser is cut down ¾ in.

Dropped and trimmed carriage

Finished floor

¾ in.

Subfloor

Nearly complete, the finish stair at left is missing only its balusters, handrail, and molding under the nosing. The newel post is mortised into the first tread for stability. This stair uses both an open stringer and treads and risers that butt-join the skirtboards or stringboards that are nailed to the wall. Above, kraft paper protects the completed oak treads from construction traffic. The open stringer is mitered to the riser, and the treads overhang the stringer by the depth of the return nosing with its scotia molding beneath. The cut-out carriage that actually supports the treads is hidden behind the finish stringer and drywall blocking.

no change will have to be made for the risers to be equal.

A more confusing condition is when the treads are thicker than the finished floor (C). At this point, I usually draw a four-riser layout, at any scale, on graph paper to figure how much of a drop I need to make, and if the bottom riser needs to be narrower.

How the cut-out carriages are attached once they are raised in the stairwell also may require adjustment at the top and bottom of the carriages (drawings, opposite page, bottom left. At the bottom, I like to use a 2x4 kickboard nailed to the floor at the front edge of the riser (A). If there is a stair opening below, the carriages can be cut to fit around the upper corner of the framing (B). Stairs take a beating, and should be well secured.

At the top, the header joist usually acts as the uppermost riser, but sometimes, the floor will extend a full or partial tread width from the framing (drawing A, opposite page, top left). A 1x4 or 2x4 ledger board can be nailed to the framing (B), and if so, the carriages must be notched to fit it. Metal angles or straps can be used if the carriages aren't exposed (C). I like a hanger board because it is quick, neat and strong. I nail the carriages to a line on a piece of plywood, a riser's distance from the top. I then raise the whole business as a unit and nail it in place (D).

Give the carriage a trial fit before sawing it out. I make only the horizontal cut that rests on the floor and the vertical cut that leans against the framing at the top before trying

the carriage in the stairwell. If you are really unsure, use a 1x10 trial board. With the carriage in place, you can easily check your layout. The treads should be level from front to back, and the carriage should fit on both sides of the opening. Also make sure that the risers will all be the same height once the treads and finish floor are installed.

If everything checks out, what's left is just cutting and fitting. With this basic layout you can produce the cut-out carriages that are needed for the stairway shown above, you can dado the carriages for let-in treads, or you can just nail cleats to the layout lines. For cut-out carriages, use a circular saw as far as you can, and finish them off with a handsaw held vertically so as not to overcut the line and weaken the cut. You can nail the triangular cutouts to a 2x6 for a third stringer if the budget is tight. Use the completed carriage as a pattern to trace onto the other 2x12, and then cut the pencil line to get an exact duplicate.

Treads and risers—On a closed stairway, the cut-out carriages sit inside the finish wall stringers, which are called skirtboards, or strings. These are usually 1x10, and should be nailed hard against the wall so that the snug fit of previously installed risers and treads isn't spoiled by the skirtboards spreading when newly scribed boards are tapped into place. They should be installed parallel to the nosing line, and as high as possible without exposing any wall where the riser and tread meet. Don't nail the cut-out carriages to the

skirtboards on a closed stairway, or the mitered stringer to the outside carriage on an open stair. Instead, hold the carriages about 3 in. away from the walls, so that they are bearing only at their tops and bottoms. This keeps the treads and risers from splitting as a result of nailing too close to their ends. Skirtboards and risers can be made of pine to ease the budget, but the best treads are oak. The standard thicknesses for treads are 1⅛ in. and 1³⁄₁₆ in.

I like to rip all of my treads and risers to width before beginning the assembly. Keep the risers a hair narrower than what's called for. Crosscut both risers and treads to 1 in. longer than the inside dimension between the skirt boards. This allows them to fit at a low enough angle to get a good scribe and still have a little extra to cut off. If the stair is open on one side, the treads will have to be rough-cut long enough to leave a ½-in. scribing allowance on the closed side, and some overhang on the open side, which gets a return nosing. The risers will need at least a 45° miter to mate with the open stringer. Use a radial arm saw or handsaw for this.

Stair assembly usually begins at the bottom. The first two risers are fit and nailed, and then the first tread is pushed tightly against the bottom edge of the riser for scribing. For the stair pictured above, I first had to cut the open, or mitered, stringer. It was pine, and was laid out like the carriages with the exception of the vertical cuts, which extend beyond by the thickness of the riser material to form

Stepladder stairway

A cleated or dadoed stairway at an angle of from 50° to 75° is considered a stepladder stairway, or ships's stairway. The rise on these ladders can be from 9 in. to 12 in. As the angle of the carriages increases, the rise increases. A 50° angle should have about a 9-in. rise. A 75° angle should have about a 12-in. rise.

As with all other stairs, the relationship between the height of the riser and the width of the tread is important. But in this case, their relationship is reversed. The tread width on stepladder stairs will always be less than the riser height.

If you know what riser height you'll be using, the easy formula for calculating the tread width is this: tread width = 20 − 4/3 riser height. If you use a 12-in. rise, then the tread will be 4 in. You can also rearrange this formula to solve for riser height (riser height = 15 − 3/4 tread width) if you know what tread width you want. This is useful if you are limited in how far out into a room the base of the stair can come.

A simple way to lay out a stepladder stairway is to lean a 2x6 or 2x8 carriage at about 75° in the stair opening. Lay a 2x4 on edge on the floor, and scribe a pencil line across its top edge to transfer a level line from the floor onto the carriage. Mark the vertical cut, at the top, in a similar manner.

Make these cuts and set the carriage back in place. Measure the floor-to-floor height and divide by 12 in. to get the number of risers. You'll end up with a whole number and a fraction. For instance, if your total rise is 106 in., the number of risers will be 8.83. That's close enough to call it 9 risers. Divide this back into the total rise of 106 in. to get an accurate riser dimension of 11.77 in., or about 11¾ in.

The easiest way to lay out the carriages is to make a story pole using dividers set at about 11¾ in. to step off nine equal segments within the 106 in. It might take a few tries adjusting the dividers to get it to come out just right. You can even find the riser dimension without the math by making a few divider runs up and down the story pole.

With the carriage in place, mark each tread from the story pole. A level line at each mark locates the treads. A bevel square set at the correct angle will also work. The treads can be cleated or let in with a dado. —B.S.

an outside miter. I'm most comfortable making these cuts with a handsaw. Keeping this angle slightly steeper than 45° allows the faces to meet in an unbroken line, without any interference at the back of the joint. The miters were predrilled, glued and nailed with 8d finish nails (photo previous page, right).

The treads are initially cut to overhang the open stringer, by the same dimension as the nosing. Then a cross-grain section is cut out so that a mitered corner is left at the outer edge. This accommodates a return nosing. I also like to put a piece of nosing at the rear of the tread where it overhangs, although you can just round off the back end of the return nosing. If the mitered stringer doesn't snug up perfectly under each tread, don't worry. The crack will be covered by the cove or scotia molding that runs under the nosing. It's more important to get good bearing for the tread on the cut-out carriage, and a little shimming or block-planing will help here.

Once beyond the open side of the stair, you will be fitting treads and risers on a closed stairway, scribing to the skirtboard on each side. With a 1-in. allowance for scribing, set the scribers at ½ in. for the first side. Set the tread or riser with the side you are going to scribe down in place on the carriage and against the skirtboard. The other end will ride high on the other skirt. If you are working on a tread, make sure it is snug to the riser along its entire length. Risers should sit firmly on the carriages to get an accurate scribe. I use a handsaw to back-bevel the cut on risers, but I keep the cut square at the front of the treads where the nosing protrudes, and then angle it the rest of the way. If necessary, use a block plane to make sure the cut fits.

Next, get the inside dimension at the back of the tread or lower edge of the riser, depending on which you are fitting. A wood ruler with a brass slide works well here. Transfer this dimension to the board you're working on, set the scriber to the remaining stock, and mark the board. Cut the boards the same way you did the first time. Cut carefully, remembering that you can always plane it off, but you can't stretch it. However, don't make the cut too strong either. Trust your measure.

Careful cutting and fitting are important with this kind of stair, and a little glue, and some nails and wedges in the right place work wonders as time goes on. Risers and treads are nailed to the carriages with 8d finish nails through predrilled holes. A sharpened de-headed 8d finish nail chucked tightly into an electric drill makes a snug hole every time.

Stairs that aren't made from rabbeted stock (like housed-stringer stairs) can be kept together if you drive three or four 6d common nails through the bottom of the riser into the back of the tread. This should be done as you fit your way up the stair. Two 1x1 blocks, 2 in. long, glued behind each step at the intersection of the upper edge of the riser and the front of the tread will cut down a lot on movement too. Any gaps between the carriage and treads should be shimmed from behind with wood wedges to eliminate squeaks. □

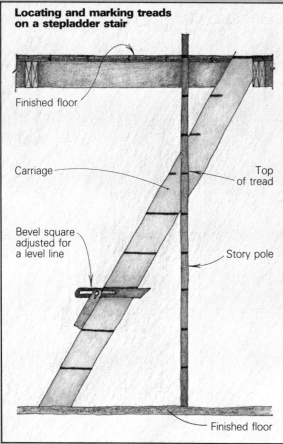

Locating and marking treads on a stepladder stair

Finished floor

Carriage

Top of tread

Bevel square adjusted for a level line

Story pole

Finished floor

Double-Helix Stair

A cherry spiral built with divine inspiration, a little steam and a lot of clamps

by Robin Ferguson

The house is majestic, sitting on a bend of Castle Creek, 8,000 ft. up in the heart of the Colorado Rockies. And a castle it seems, with its medieval lines and three towers. The central, and highest, tower was designed for solitude, meditation and glimpsing the sunset. When the clients looked at the architect's model of the house, they liked the tower. It was just what they wanted, but with the tower centered over the living room, they wondered how they'd get up there. In response, Steven Conger and Michael Martins, who had designed the house, conceived a sculptural, double-helix stair (photo right).

The design and construction of the stair was a group effort that evolved over many months. On hearing the architects' concept for the stair, construction foreman Ivar Eidsmo said he'd seen one just like what they had in mind. He was referring to the Miraculous Staircase in the Loretto Chapel in Santa Fe, N. Mex. (photo below). It was built about 1878 by an itinerant carpenter who appeared out of nowhere, apparently in re-

Unlike conventional spiral stairs that are supported by a central post, the cherry staircase at right stands on two helical stringers. It rises over 16 ft. to an observation tower above the living room. The design of this double-helix stair was inspired by the Miraculous Staircase (above), which leads to the choir loft of the Loretto Chapel in Santa Fe, N. Mex.

Alan Becker

From *Fine Homebuilding* magazine (October 1987) 42:45-49

Plan view of column and cage

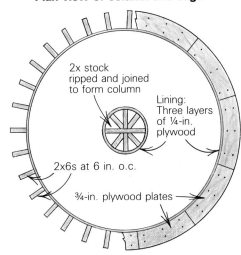

2x stock ripped and joined to form column

Lining: Three layers of ¼-in. plywood

2x6s at 6 in. o.c.

¾-in. plywood plates

sponse to prayers to St. Joseph, carpenter of Galilee. The anonymous carpenter worked four months and disappeared as mysteriously as he had arrived, without finishing the railing and without payment for his work.

Eidsmo went down to Santa Fe and photographed the Miraculous Staircase. Although the stair we built is very different structurally and is shorter, narrower and much less ambitious, the Miraculous Staircase did serve as inspiration.

Rising over 16 ft. around a 2½-ft. radius and revolving 1½ turns along the way, our stair was a major undertaking. It was begun by Chuck Miller, one of the carpenters working on the house. I came along later, after the layout had been done and the forms built. Initially my job was to speed up production, but when Miller had to leave the project, I took responsibility for its completion.

Specifications—During the design stage, an engineering firm, Nicol and Giltner, was hired to do structural specifications for the stair. To match the interior woodwork, the staircase is made of cherry. So the engineers based their calculations on the strength of that wood.

They determined that the stringers should be 2 in. thick, composed of eight layers: four ¼-in. plies in the middle, two ⁵⁄₁₆-in. crossband plies with their grain running perpendicular to the others, and two ³⁄₁₆-in. face veneers.

Because of the wedge-shaped treads, the inside stringer needed to be 7 in. wide, while the outer one had to be 14 in. wide. At the top and bottom of each stringer, a ¼-in. steel plate had to be sandwiched across the full width and securely lag-bolted to the landing and floor system. To hide these plates and cover the various structural laminations, we had to glue 1-in. thick caps, made with eight layers of solid cherry, to the upper and lower edges of the stringers.

In addition, the engineering firm also specified the 2-in. thickness of the treads and their installation details. The treads are held in place by five ½-in. dia. lag bolts 6 in. long—two through the inside stringer and three through the

outside one. The heads of the bolts are concealed beneath the final ply.

Since the stringers are so long (the outside one is nearly 28 ft.), each ply is composed of sections, 6 ft. to 8 ft. long, glued up to form one full-length piece. For maximum strength, the adjoining ends of each ply were scarf-jointed (drawing, top left) on a huge industrial table saw at a local millwork shop.

Column and cage—Building the laminating forms was the first big hurdle in the construction process. Rather than build the stair in a woodshop and then transport and install it, we set up the laminating forms right in the stairwell. The stringers were laminated in place, and the forms were then stripped away from them.

We built an inner column of solid 2x fir, and an outer cage encircling it (drawing, middle left). These we wrapped with plywood along the path of the stringers. The cage consisted of vertical 2x6s on 6-in. centers. We used ¾-in. plywood to make the curved plates by bandsawing segments to form a 5-ft. dia. circle.

The column was made with seven pieces of 2x stock stood on end and cut to radiate out from a central axis, like spokes on a wheel, to create a star-shaped cross section with a 12-in. diameter. Three layers of ¼-in. AC plywood were used for the lining, and this is where our wood-bending difficulties began.

Straight curves are relatively easy to bend, but compound or twisting curves put a much greater stress on the wood, so we had to use steam. Eventually we built three different steamboxes, each one longer than the last. These were simple plywood boxes (6 in. by 14 in.) with a perforated copper pipe running through them, and fed by a 2½-gal. kettle (photo below left). Steaming made the plies supple enough, but once out of the box, the thin wood cooled so fast that we had only seconds to get the piece bent in the form. With the help of several people, we held it there and screwed it to the form with plenty of drywall screws.

As each full-length layer was completed, careful truing had to be done before the next could be added. We checked the curves with templates cut to the appropriate radius (which changed with each layer). Any high spots caused by the steaming or by slight imperfections in the column or the 2x6 cage were removed with a lot of grinding and hand sanding. This was a tedious process, and given the length of the forms and the cramped 2-ft. space between column and cage where we were working, we came to appreciate the term "cage."

Clamping setup—The engineering firm we consulted had specified resorcinol as the best glue for our project. Commonly used in plywood and gluelam beams, resorcinol requires extreme clamping pressure (175 psi) to bond properly. To satisfy this requirement, we used vertical 2x6s on edge, forming a continuous row of cauls or battens along the length of the outside stringer. We ran ⅜-in. threaded rods through the ends of the cauls and into steel plates that had been tapped and lagged securely to the form (drawing, facing page, top right). Now we

The twisting curves in this stair are so severe that even the thin laminations had to be steamed before they were pliable enough to bend around the forms.

Photo: Germaine Dietch

could apply plenty of pressure and distribute that pressure uniformly.

Before starting on the stringers, we laminated a pressure board from four layers of ¼-in. plywood to place over the stringer during the gluing operation to protect it and further distribute the clamping force of the cauls.

Before we could begin laminating stringers, we built another steambox, this one 30 ft. long, to accommodate the veneers. Once we got started, we soon learned that oversteaming did no good. After about 20 minutes, the cherry veneers were as pliable as they were going to get. Extra time only made the wood swell—a potential disaster when it shrank later.

As with the plywood, the cherry plies cooled rapidly, and getting the pieces from the box to the form in time was hard. Four people would run the plies into the house and feed them up to four others positioned around the form. Cherry isn't the best wood for bending, but there could be no dry runs—glue was rolled on to the previous lamination, the ply run in and clamped.

Resorcinol sets in about two hours—so the directions say—but here in the Rockies, the humidity is very low. Getting the ply in place, pressure boards on and all 90-odd clamps tightened before the glue set was frantic work.

On the inside stringer, we couldn't use the same clamping technique because the radius was too tight and the rise too steep. Instead of full-length plies, we laid up sections and clamped them with metal straps and a Signode model DO-3A banding machine (Signode Corp., 3610 W. Lake Ave., Glenview, Ill. 60025), like those used in lumberyards to bind stacks of plywood. Using about 30 bands around the column for a 5-ft. length, we achieved tremendous pressure. This method worked very well. As with the form itself, each full-length layer was carefully trued before the next was applied.

Adding the caps—We could laminate only six of the eight layers around the form. So after laminating as many as we could, our next step was to square the upper and lower edges of the stringers and glue on the caps. For this we used a 1½-in. straight-flute bit in a router. To keep the bit from wobbling on the curved surface, we had to sculpt wooden bases for the router, both convex and concave, to match the stringer's shape.

We attached a flexible guide strip along the face of the stringer for the router to ride against. As with most of the work on this stair, there was no convenient way to arrange a trial run. So the routing itself was a very intense process; we were keenly aware that an errant cut couldn't be easily hidden or fixed.

To apply the cap strips, we had to rethink our clamping methods, for pressure was now required in two directions: against the form and onto the stringer. We drilled 3-in. thick cherry blocks and threaded bolts through them. With the blocks screwed to the form, the bolts could be run down to exert pressure on the cap (drawing, bottom right). Even though we were using machine-thread bolts, the wooden threads they cut in the hardwood blocks were plenty strong enough for our purpose.

For holding the cap strips against the form,

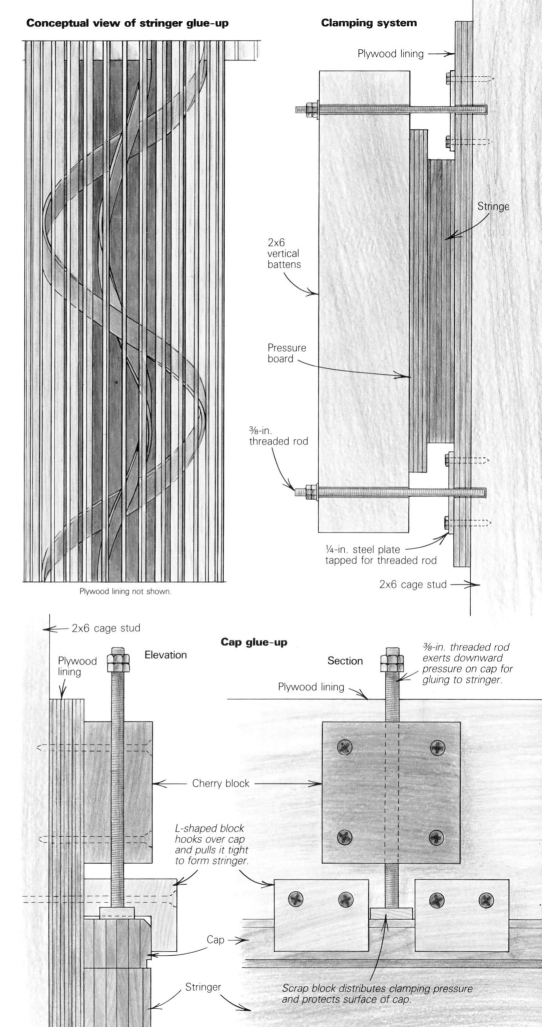

Conceptual view of stringer glue-up

Plywood lining not shown.

Clamping system

Plywood lining

2x6 vertical battens

Pressure board

⅜-in. threaded rod

Stringer

¼-in. steel plate tapped for threaded rod

2x6 cage stud

Cap glue-up

2x6 cage stud

Plywood lining

Elevation

Section

⅜-in. threaded rod exerts downward pressure on cap for gluing to stringer.

Plywood lining

Cherry block

L-shaped block hooks over cap and pulls it tight to form stringer.

Cap

Stringer

Scrap block distributes clamping pressure and protects surface of cap.

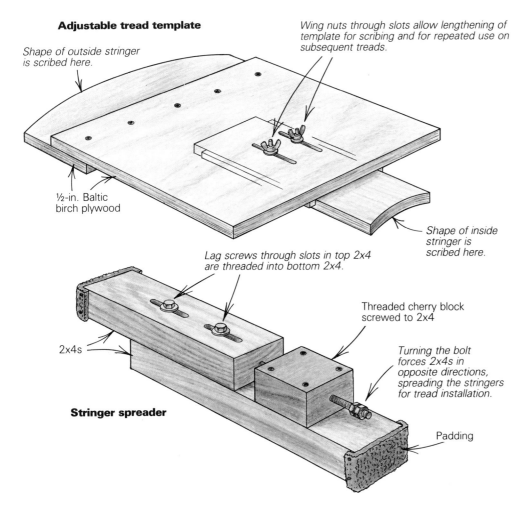

Adjustable tread template

Shape of outside stringer is scribed here.

Wing nuts through slots allow lengthening of template for scribing and for repeated use on subsequent treads.

½-in. Baltic birch plywood

Shape of inside stringer is scribed here.

Lag screws through slots in top 2x4 are threaded into bottom 2x4.

Threaded cherry block screwed to 2x4

2x4s

Turning the bolt forces 2x4s in opposite directions, spreading the stringers for tread installation.

Stringer spreader

Padding

Truck-tire inner tubes, cut up and laced into long bands, were used to clamp the final layer of cherry in place while the epoxy set.

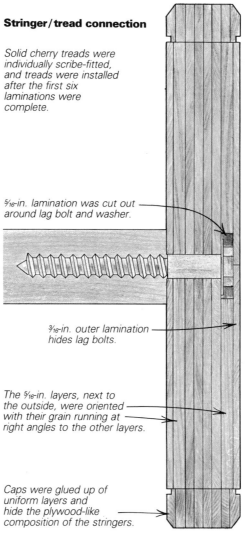

Stringer/tread connection

Solid cherry treads were individually scribe-fitted, and treads were installed after the first six laminations were complete.

⁵⁄₁₆-in. lamination was cut out around lag bolt and washer.

³⁄₁₆-in. outer lamination hides lag bolts.

The ⁵⁄₁₆-in. layers, next to the outside, were oriented with their grain running at right angles to the other layers.

Caps were glued up of uniform layers and hide the plywood-like composition of the stringers.

we cut 2-in. wide L-shaped blocks that hooked over the edge of the cap. We spaced them every 4 in. and pulled them tight to the form with drywall screws. Once the cap strips were secure, we no longer needed the forms and set about to dismantle them. All the drywall screws did their best to keep the column and cage intact, and it seemed for a time that the stringers would be destroyed with the forms.

Treads—The main purpose of all the careful truing was so that one tread pattern could be used for the whole stair. But so much for careful planning. With the forms stripped away, the stringers took a set of their own, and while the differences were minute, a single tread pattern was now out of the question. We were obliged to cut 21 individual treads, each having to fit perfectly between the convex and concave surfaces of the stringers.

In order to make individual patterns for each of the treads, I built an adjustable template of ½-in. Baltic birch plywood (drawing, top left). I made it in two pieces, held together with bolts and wing nuts in oblong slots, and cut each end to the appropriate radius. It could then be extended to make contact with both stringers. After scribing the exact curve, I belt-sanded down to my line. I repeated the process (each time opening the jig slightly) until I had a perfect fit. The template was then used to make a pattern of ⅛-in. tempered hardboard. We glued each pattern to a tread blank, bandsawed to within ⅛-in., and trimmed it flush with a 2-in. bearing-over router bit.

The treads were such a tight fit that they couldn't be installed without marring the faces of the stringers. We had to make another device to spread the stringers slightly so we could get them into place (drawing, middle left). For this I used two pieces of 2x4, held together with lag bolts run through slots, and just tight enough to hold the two together yet allow movement. With one of the blocks of cherry from the cap glue-up (tapped for a bolt) mounted on one of the 2x4s, and the bolt against the end of the other, the two could be forced apart by turning the bolt.

Having installed all 21 treads, we were able to use the stair. It was quite a sensation, after so many months, finally being able to climb the thing. Even so, some of the most challenging work still lay ahead.

Epoxy to the rescue—Applying the last two plies required still other methods of clamping. Since we had dismantled the forms, our old system was no longer available to us. Satisfying the high pressure requirements for resorcinol would have dictated major retooling.

I knew about epoxies but had never used them for anything but plastic laminates. But for the final plies they proved a great problem solver. Instead of resorcinol, we used epoxy that had an open time of six hours, with a full cure in seven days. Rather than clamp each ⁵⁄₁₆-in. layer, we stapled it in place, since contact was all the epoxy needed for a strong bond. Before clamping, we cut holes in the wood around the lag bolts, and after the epoxy had set, ground the heads flush. These holes in turn were filled with

Photo: Michael Owsley

auto-body filler and sanded smooth. It was time at last for the face veneer of cherry.

We couldn't use staples on the last veneer so our clamping system changed again. To hold the veneer tight to the stringer during glue-up, we used a series of 2x2 vertical cauls, snugged against the veneer with 30 truck-tire inner tubes. After cutting them into rings and lacing them into bands, we wove the tubing between the treads and around the opposite stringer (photo facing page). This worked amazingly well.

Railing and connections—The handrails had been glued up before the forms were dismantled, but they needed their upper and lower surfaces squared before they could be installed. Trying to work the outside rail required a great deal of patience. At nearly 28 ft. long, 1½ turns and a 5-ft. diameter (photo top right), just getting it clamped was like wrestling a giant snake. Once it was clamped, we could work only a short portion before the twist had us working upside-down and we were forced to reposition it.

We hand-planed the surfaces of the rail, trying to follow the twist with each stroke. It was impossible to avoid some tearout because, with all the plies, there was always grain running in opposite directions. But we cleaned this up with sharp cabinet scrapers.

Once shaped, the rails were clamped temporarily in place. Because of slight variations in the distance between stringer and rail, the square balusters had to be fit individually. They are doweled into both the handrail and stringer.

One unusual feature of this stair is that it has two handrails. A regular spiral stair built with a center post has only an outside handrail. Our stair also has a rail that hovers just above the inside stringer, so you can hold on with both hands as you go up or down. Two rails make the stair more comfortable and secure to use.

Foam patterns—The final phase, and for me the most enjoyable, was designing and carving the many rail connections. At the suggestion of Michael Owsley, who aided greatly in the completion of this stair, we glued up pieces of polystyrene foam into big chunks. We wanted to experiment with these before making the transitional rail connections in cherry.

After permanently installing all the long sections of rail, we glued the polystyrene chunks in place between the rails with five-minute epoxy and a piece of paper sandwiched between the foam and wood. (The paper allowed us to break apart the glue joints later.) These were then shaped with various saws and rasps to form the final models (photo bottom right).

Passing inspection—The Uniform Building Code says that a staircase shall not rise more than 12 ft. vertically between landings. Our staircase rises over 16 ft., but we were granted an official variance by the building department. Since this is a low-use stair that leads only to an observation tower and is not for egress, the inspector felt the variance was justified. ☐

Robin Ferguson is an architectural woodworker in Snowmass, Colo.

Maneuvering the 28-ft. long handrail into place was like wrestling a giant snake (top). Once it was installed, the builders glued foam blocks in place temporarily (above) to work out the shapes for the transitional sections of handrail.

Traditional Stairways Off the Shelf

About manufactured stair parts, and how they are made

by Kevin Ireton

On the banks of the Fox River in Oshkosh, Wis., stands an old stone building with a Victorian staircase just inside the front door. It's a heavy stair, done in dark oak, with a balustrade of spindles ascending on the left, complemented by a raised-panel wainscot on the right. The box newel that anchors the balustrade is also made of raised panels and is topped off by a square cap with a graceful curving cross section. The newel cap isn't nailed to the post; if you lift up on it, it comes off easily. Underneath you'll find a small scrap of sandpaper with writing on the back. In a faded penciled scrawl it says: "Built by R. W. Maurice Jan. 18, 1897." Such is the nature of stairbuilding and the pride it generates; people sign their work.

Stairbuilding is the pinnacle of the carpenter's trade. It combines the mathematical complexity

of roof framing with the exacting standards of furniture-quality finish work. Traditionally, the staircase is the dominant architectural feature inside a house.

From the end of World War II through the early 1970s, during the heyday of the rambling ranch-style house, the practice of stairbuilding floundered, not only because the indiscriminate suburban sprawl encouraged one-story and split-level houses, but also because mass production and modern technology were streamlining construction techniques wherever possible—and one result was that wrought iron became the

A great deal of handwork goes into manufactured stair parts. Above, a worker at Morgan Products Ltd. uses a chisel to carve the inside of a volute where the shaper couldn't reach.

material of choice for many stair railings. But rising land costs have meant a return to multi-story houses. Also, the current popularity of old-house renovation has brought with it the need to restore and rebuild the stairwork of 18th and 19th-century craftsmen. All of this has fueled a resurgence of interest in traditional stairbuilding.

But the resurgence has been hamstrung by a lack of carpenters with an extensive knowledge of stairwork. Without an experienced stairbuilder on his crew, a builder has two choices. He can either contract with a custom-stair outfit to build the stairway to specifications, or he can design the stairway around manufactured parts. In a future article, I'll write about custom stairbuilders, but here I'll deal with how and where off-the-shelf stair parts are manufactured and sold. To find out about these exotic-looking

From *Fine Homebuilding* magazine (December 1986) 36:56-61

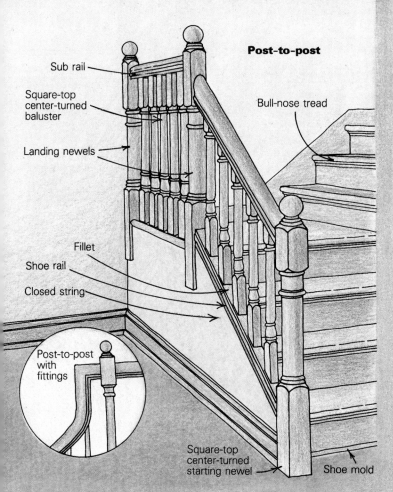

Post-to-post

Sub rail

Square-top center-turned baluster

Landing newels

Fillet

Shoe rail

Closed string

Bull-nose tread

Post-to-post with fittings

Square-top center-turned starting newel

Shoe mold

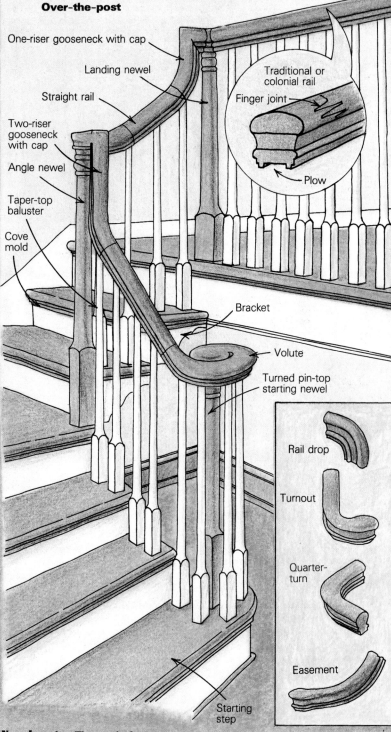

Over-the-post

One-riser gooseneck with cap

Landing newel

Straight rail

Two-riser gooseneck with cap

Angle newel

Taper-top baluster

Cove mold

Traditional or colonial rail

Finger joint

Plow

Bracket

Volute

Turned pin-top starting newel

Rail drop

Turnout

Quarter-turn

Easement

Starting step

Stairbuilder's lexicon

Balusters—Vertical members used to support the handrail and to fill the open area between the handrail and the tread or floor. *Taper-top* balusters (above right) are turnings whose shaft tapers, usually from 1¼ in. in dia. to ¾ in. at the top. *Square-top* balusters (above left) are used with plowed (grooved) handrail and can be uniformly square in cross section or *center-turned*, which means that the center section is a decorative turning.

Balustrade—The complete railing system, including newel posts, balusters and handrail. There are two basic types: in a *post-to-post* system (above left), the handrail is not continuous but is fitted between newel posts; an *over-the-post system* (above right) uses handrail fittings to create a continuous flow of handrail over the newel posts.

Brackets—Thin decorative pieces that are attached under the returned nosing of an open-string stair.

Closed string—A solid stringer that covers the ends of the treads and risers such that their profile cannot be seen (above left).

Fillet—A thin strip that fills the plowed (grooved) space between balusters in a handrail, sub-rail, or shoe rail.

Fittings—Sections of handrail used at the beginning and end of a balustrade or wherever the handrail changes height or direction. An *easement* is a fitting that curves in a vertical plane, used to change the angle of the handrail (shown at right is an easement with an integral newel cap). A *gooseneck* (above right) includes an easement and is used to change the handrail from the incline of the stair back to level, either at a landing or at the top of the stair.

A *quarter-turn* (right) is a level section of handrail used to make a right-angle turn. A *rail drop* is a curved fitting used as a decorative beginning or ending on a handrail (right). A *turnout* (right) is a starting fitting that curves in a level plane before curving vertically up the stair. A *volute* (right), or a wreath, is a starting fitting, similar to a turnout, that scrolls into a tight circle.

Handrail—The horizontal or inclined member that runs over the balusters and is supported by the newel posts. *Plowed* handrail (see detail above right) has a wide groove cut in its underside and is used with square-top balusters. *Wall rail*, usually simpler in design than handrail, is used where a stair runs along a wall and is attached to the wall with brackets.

Newel posts—The vertical members, larger than balusters, that support the balustrade. The *angle newel* (shown above) is the longest of the newels and is used at a landing where a stair changes directions and then continues to climb. A *center-turned newel* (above left) has *square ends* but a lathe turning in the center. A *box newel* is rectangular in cross section its entire length. A *landing newel* (shown above) is shorter than an angle newel and is used at a landing or at the top of a stair where the handrail changes from inclined to level. A *pin-top newel* has a dowel-like pin turned at its top, used to attach a fitting. A *starting newel* is simply the first newel of the balustrade.

Open string—A stringer that is cut out for the treads and risers such that their profile can be seen from the side (above right).

Shoe rail—A plowed rail that is used under square-end balusters when they don't sit directly on the treads (above left).

Starting step—The first tread and riser at the bottom of a stair, usually a step that is curved on one or both ends (above right).

Sub-rail—A thin rail attached to the underside of a handrail to make it more massive and allow for the use of square-top balusters (shown above left).

components, I visited some of the companies that make them.

Around the country, there are a handful of companies whose volume of sales qualifies them as major manufacturers of stair parts (see the sidebar on p. 25). I was able to visit three of them and tour their plants. I learned that these companies are highly competitive (at the distributor level), protective of their secrets and fiercely proud of their products. I also learned that, much to each other's chagrin, all three produce high-quality stair parts.

Stair-part basics—The balustrade of a staircase is the complete railing system, including newel posts, balusters and handrail. There are two basic types of rail systems—a post-to-post system uses square-topped or box newels with straight sections of handrail fitted between them, and an over-the-post system uses turned pin-top newels and various fittings—volutes, easements, goosenecks and quarter-turns—that allow a continuous flow of handrail (for definitions of the individual stair parts, see the previous page).

Most manufactured stair parts relate to the balustrade, but all the companies produce various other components as well. For instance, they all offer several styles and sizes of starting steps with curved risers. These are used with over-the-post balustrades that begin with a volute, and in some cases with a turnout. They serve as a structural and visual anchor for the balustrade, the shape of the starting tread reflecting the shape of the handrail.

Some companies make bull-nose oak treads for the rest of the stair, and these are available in various widths and lengths. You can also get nosings, cove and shoe mold, all to match the treads. Most of the companies offer decorative bandsawn brackets for use on the side of an open stringer stair. A few companies even make tread and riser caps for use along the outside edge of a stair that's to be carpeted. These give the illusion of a finished stair but save the cost of oak treads and risers that would otherwise be covered by the carpet.

Most of the industry, however, is devoted to producing newel posts, balusters, handrails and handrail fittings. The designs for many of these parts are very old, derived from the relatively delicate balustrades of colonial America. But some of the companies have recently come out with lines of stair parts that are more massive, including handrails as wide as 3½ in.

The three companies that I visited all have different product lines, though the most popular styles are available from all of them. They sell only through distributors, so if you contact them directly, they'll send you a catalog and the name of their distributor in your area.

Morgan—My first trip was to Morgan Products Ltd. in Oshkosh, Wis., where I met Sam Stecker, product manager for Morgan's line of stair parts. It was Stecker who showed me, with some ceremony, the signed note under the newel cap in Morgan's office building. After talking for an hour in his office, we walked outside into bright sunshine and crossed the street to the plant.

Morgan has been in business since 1855, when it operated a planing mill on the same site that it occupies today. But now they're spread out over 27 acres and the plant is a labyrinthine conglomeration of buildings connected by walkways, conveyor belts, railroad tracks and even dirt paths. Some of the buildings are fairly new; others predate the 1897 note under the newel cap.

The stair parts that Morgan produces are either oak or birch, except for a small line of pre-assembled railings that are done in hemlock. They buy their lumber roughsawn, mostly from mills in the upper Midwest and Canada. Most of it is dried in their own kilns, a process that takes from 30 to 45 days depending on species, size and initial moisture content. Then the lumber enters the plant and is run through surface planers that give the operators their first glimpse of any defects. The next stages are crosscutting and ripping.

"Any manufacturer will tell you that a sawyer can make or break them," Stecker told me. The sawyers at Morgan use saws that rise up into the stock from beneath the tables. The planed and ripped pieces come to them on conveyor belts. The sawyers scan each piece for knots or other defects and gauge the distance between them. Next they slide the piece along under the saw guard, butting its end into one of a series of fixed stops on the other side of the saw. There are maybe a dozen of these stops, each for a specific stair part—baluster, newel or tread. The sawyer's job is to cut out all the knots, but at the same time, to get the longest possible piece.

From here the pieces for newels and fittings go on to gluing operations, where machines spread them with urea-base glue, clamp them and heat them at the same time to speed the curing process. The glued-up blanks are ready to be worked half an hour later. Pieces for turned newel posts and balusters are planed again to remove the excess glue and bring them down to final dimensions. Then these turning blanks are moved on carts to the automated back-knife lathes. Some of the baluster lathes are hand fed by an operator, but most are self feeding, holding up to 12 blanks at a time.

There are two basic types of balusters. Taper-top balusters are available with various amounts of decorative turning, but their distinguishing characteristic is that most of the shaft is a straight taper from about 1¼-in. dia. at the bottom to ¾-in. dia. near the top. Actually, the taper stops just below the top, and the last 3 in. is a uniform diameter. This is so that cutting the balusters to length doesn't affect the size of drill bit needed to install them in the handrail.

The other type of baluster is called square top or square end, and is used with a plowed handrail. The balusters are cut on the rake angle of the stair and toenailed into the underside of the handrail. The plowed areas between the balusters are filled in with thin strips of wood called fillets. The square-top balusters that Morgan offers are all center turned, which means that both ends are square but the center section has a decorative lathe-turned pattern cut into it.

When Stecker and I arrived at the station where newels are turned, the lathe operator was changing knives for a new run. He said the whole process took anywhere from two to four hours. But once it was set up, he could turn out about 200 newels a day.

All the turnings go on to sanding machines where they get spun again, this time against rows of sanding belts of various widths (top left photo, facing page). These belts are backed by heavy broomstraws that press them against the baluster or newel.

Before being glued up into blanks, the stock for handrails is finger-jointed end to end to obtain the required length. I have worked with manufactured handrails and had always objected to the finger joints. Though strong and well done, they detract from the visual integrity of the handrail, especially when the adjoining pieces are markedly different colors.

I asked Stecker why they manufactured finger-jointed handrails. "We cut for yield," he told me. "It's a high-volume business. We manufacture miles and miles of handrail every week, and there just aren't enough clear lengths of lumber out there in the new-growth trees." They're also concerned about straightness and stability. Long pieces of solid handrail are more liable to twist or warp than glued-up rails.

Another advantage of finger joints and laminations is that Morgan can use up wood that might otherwise be wasted. This economizing is evident all over the factory. Everything is used for stair parts that possibly can be, most of what can't be used is chopped up and burned, either to run the kilns or to heat the plant.

Morgan offers four styles of handrail, with various other options possible using a sub-rail. However, their fittings are available in only one style. The most popular handrail—at Morgan and elsewhere—is called colonial by some companies and traditional by others (drawing, p. 21, inset at right). Morgan's traditional rail, which they call M-720, is stack-laminated with three finger-jointed layers ⅞ in. thick. Individual pieces in each layer are seldom shorter than 24 in., and the top layer of the handrail is limited to three or four finger joints, depending on its length.

After the handrail blanks are glued up, they're fed into huge molders that turn out smooth finished rails in one pass. No sanding is needed.

I asked Stecker how quality control was maintained. "We have 70 quality-control inspectors for stair parts alone. All our employees have the authority to reject a piece of wood that comes to their station. No questions asked." And sure enough, at each station there was a small pile of rejects from the operator. Some of these get chopped up, some are used by the salespeople in their training sessions. Still others turn up around the factory. Defective newels become the corner posts on the flatbed carts used to move stock from one station to another. In several places, I saw tables with balusters for legs. When I visited the product-development offices, I found rejected newel caps with holes drilled in them doing duty as pen and pencil holders.

The carpenter shop—In the course of the tour, Stecker and I had worked our way toward the carpenter shop. Straight handrail, balusters and turned newels are all pretty much the products of automation; the operators only feed and tend the machines. But all of the fittings—vo-

Morgan's taper-top balusters are sanded on the machine above. They are spun at high speed against rows of sanding belts at the back of the machine. The belts are made in various widths to conform to the shape of the balusters and are forced against the baluster by stiff broomstraws.

Handrail fittings look more complicated to make than they really are. They all begin as simple shapes cut out on the bandsaw. The Morgan operator in the photo above is making level quarter-turns for use around a stairwell opening.

At Visador/Coffman, the shape of the volute allows the operator to pass it all the way around the shaper, thereby avoiding a great deal of handwork. In the photo below, he is using a template that holds two volutes at a time, allowing him to shape the inside curve of one and the outside curve of the other. He will make the final pass on the shaper head to his left.

Installing Manufactured Stair Parts

Laying the handrail on the stairs is the first step

by Sebastian Eggert

In every cowboy movie there's a barroom brawl where someone crashes through the railing at the top of the stairs. If this could happen as easily as it looks, then our building codes would have outlawed wooden railings long ago. In reality, a well designed and built balustrade is not only beautiful, but it's also strong.

Like most builders today, I use stock parts for much of my stairwork. I buy the treads, tread-return nosing, newel posts, balusters and handrail over the counter (see pp. 20-25). All that's left is the challenge of installing them.

There are two types of balustrades that you can build. A post-to-post system has sections of handrail, either straight or with fittings, that fit between newel posts. This system is easier to install than the other. But for all the kids who enjoy sliding down the banister, it'll never do.

An over-the-post system uses straight rail joined to fittings, and flows gracefully over the tops of the newel posts (photo below left). This is the type of balustrade I usually build, and it's the one I'll discuss here.

The rough stair—Whenever I can, I frame the rough stair myself, rather than installing finished treads and risers over someone else's framing. This way I'm sure that the stairs will meet code requirements, and I avoid problems with carelessly installed carriages. At the same time, I can add blocking—for newel posts and wall-rail brackets—that I'll need later. (For more on stair construction, see pp. 8-14.)

After roughing in the stairs, I pull off the job until most of the other work is done. I like to get back just before the carpeting goes down. Since the stairs are often the only way to move between floors, I work the second shift if possible, to avoid interruptions by other tradespeople working on the house.

Lay it out in place—Some installation instructions included with stair parts tell you just enough to get you into trouble. In particular, the tables that list the heights of newel posts can be misleading. They work fine if your staircase is built just like the diagrams—same rise/run, same handrail heights, and most important, the same newel-post locations. But every installation is different, and most have their quirks.

The best way to build a balustrade is to lay out the handrail, fittings and all, in its exact location, right on top of the stair treads, the same way you lay out the plates of a stud wall (photo below right). Laid on the stairs, the handrail should touch the nose of each tread. The newel-cap fittings should sit directly above the centers of the newel-post locations.

Be aware that there is a procedural catch in the assembly process. The handrails need to be laid out over the locations of the newel posts,

When laying out a balustrade, measuring the components in place works best. Clamp the handrail and fittings in place right on top of the treads to determine where and at what angle to cut them. The finished installation shows a starting newel with a turnout and a two-riser gooseneck that had to be extended because of the location of the landing newel.

and those locations cannot be found easily without the treads in place. But the treads cannot be permanently installed until the risers and mitered stringers are on, and in some cases they can't go on until the newel posts are in. However, if the newel posts are in, you can't lay out the handrail on the treads. To cope with this confusion, I don't install anything permanently until the handrails are assembled.

Newel-post layout—Newel posts are the foundation of any handrail, and their placement is critical. Always plan for a newel post at the top and bottom of a run of stairs, and any place there is a landing or change of direction. Runs of over 10 ft. should have an intermediate newel post to strengthen them.

Except for starting newels with volutes and turnouts, which stand off to the side, newel posts should be laid out on the centerline of the balusters and handrail. Starting newels are usually notched around the corner of the first step. Newel posts at a landing or at the top of the stairs sit on the intersection of baluster centerlines or where dictated by the configuration of the fitting being used.

Turnouts and volutes sit on newel posts that are fastened to starting steps, which extend out from the staircase edge in a semicircle, with the newel post at the center of the circle. The manufacturers supply paper templates with these fittings to show the newel-post and baluster locations on the starting step. With all the newel-post locations marked, you can begin assembling the handrail.

Cutting the fittings—Handrail fittings are used to change the direction or slope of the handrail. Those used to change the slope include a curved piece called an easing or easement that has to be cut for the particular angle (the rake) of the staircase where it's being used.

I use a pitch block to mark the cuts on the fittings. A pitch block is a triangular piece, usually of wood, whose sides represent the rise, run and rake angle of the stair. One of the pieces cut out of the rough stair carriage will serve the purpose, or you can make a new pitch block from a scrap of 2x stock with the same dimensions as the rise and run of the stairway. If you're working on a stair with a landing and you didn't frame it yourself, check that the two flights of stairs have exactly the same rise and run. If they don't, make a separate pitch block for each flight to ensure accuracy.

At the bottom of the stairs, the handrail begins with a starting easing, turnout or volute. I set this fitting on a flat surface and snug the pitch block (run side down) to the underside where the easing turns up. Then I mark the point (tangent) where the hypotenuse of the pitch block touches the curve of the fitting (top photo at right). Then I turn the pitch block over on its short leg (rise side down) and scribe a line on the fitting (along the rake side of the pitch block) that passes through the first mark I made (photo at right, second from top). This gives the angle at which to cut the fitting.

It's hard to hold the fittings securely in a miter box, and at the proper angle to get an accurate

cut. Sometimes it helps to use the pitch block on the miter-box table to hold the fitting at the right angle. Fittings that have an integral newel cap can't be laid squarely against the fence, so for them I screw a piece of plywood to the bottom that acts as a jig (photo bottom right). With some fittings, like up easings, it's easier just to hold them by hand and hope for the best.

I use a Teflon-coated 80-tooth carbide crosscut blade to make mirror-smooth cuts. Cut just shy of the line, and if the angle looks good, go for the final slice. Take too much and you've got a $50 piece of kindling.

To test the cut, clamp to the stairs a section of straight rail that has a square cut on one end, and while holding the newly cut fitting against the square cut, check the underside of the fitting with a torpedo level. If it reads between the lines, you're alright. Don't try to use your level on the top of the fitting because the millwork is rounded and not consistent enough to give a true reading.

Rail bolts—The next step is to join the fitting to a section of straight rail long enough to reach the next fitting location. I use rail bolts for the connections. These are 3½-in. double-ended bolts, with a machine-screw thread on one end and a lag-screw thread on the other. To locate the holes for the rail bolt, cut a wafer-thin piece of handrail (about 3/16 in.) to use as a template. Drill a small hole through the template on the vertical centerline 15/16 in. from the bottom of the handrail. Match up the template with the adjacent ends of the rail and fitting, and mark the hole locations with your pencil.

Drill a ¼-in. hole 2 in. deep in the end of the fitting and turn the lag-screw end of the rail bolt into it. To do this, I spin two hex nuts on the machine-screw end and lock them against each other, then use a wrench to turn the bolt into the fitting (photo at right, third from top). You can also just clamp the rail bolt with a pair of vise grips and turn it that way, but be careful not to damage the machine-screw threads.

Next, drill a 3/8-in. dia. hole at least 1 in. deep in the end of the straight rail. Then mark the bottom of the handrail 1 3/8 in. from the end and on the centerline, and drill a 1-in. dia. hole 1½ in. deep. Be careful not to drill too deep or the point of the bit will come out the top of the handrail—another expensive mistake.

This last hole is the cavity where the nut is turned on the rail bolt's machine threads to pull the fitting and the straight rail together. After final assembly, you cover the hole with a wooden plug supplied with the rail bolt. I square up a portion of the hole facing the joint with a ½-in. chisel to provide a flat surface for the washer and nut to bear against.

Check the alignment of the fitting to the rail. If it's off, ream the 3/8-in. hole just enough to line the two up. Don't expect a perfect match. The cross sections of straight rail and fittings are always slightly different, so try to line up the bottoms and the side profiles as much as possible.

Most companies supply the rail bolts with star-shaped nuts that you turn with a hammer and a nail set. But the tapping upsets the alignment and levering mars the sides of the hole, so

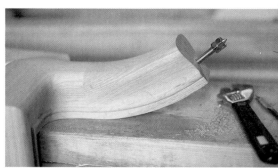

These photos show a gooseneck fitting with a newel cap, quarter turn and an up easing. In the top photo the pitch block is being used (run side down) to locate where to cut the up easing. This procedure is the same as for any starting fitting, since at this point the up easing is the start of a new flight of stairs. In the next photo, the pitch block has been flipped over (rise side down) to mark the angle at which to cut the up easing. The photo above shows the fitting after it has been cut and with a rail bolt attached. The two hex nuts on the rail bolt were locked against each other and used to turn the rail bolt into the up easing.

Wherever the shape of the fitting makes it difficult to maneuver in the miter box, the author screws a piece of plywood to the bottom of the fitting as a jig. Here he's cutting a turnout easing with newel cap.

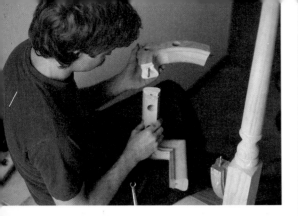

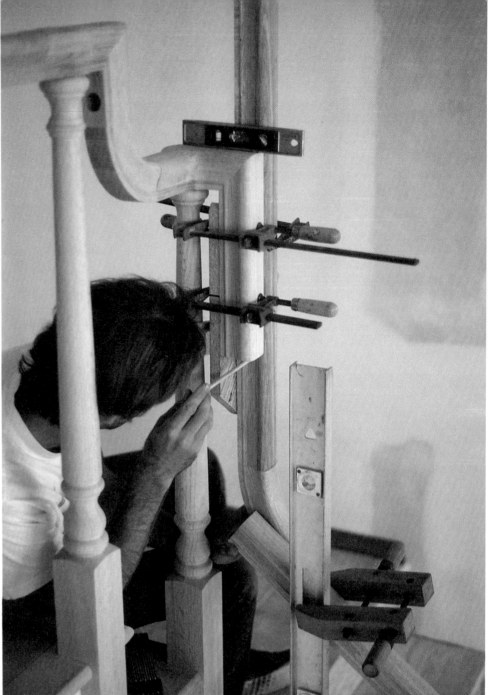

Ready for final assembly, the joining faces are coated with glue, the rail bolt inserted into the adjacent fitting (top left), and the nut and washer started, then tightened, through the large hole on the bottom of the fitting (middle left). Because they make for a smoother assembly, the author prefers hex nuts and a box wrench to the star-shaped nuts supplied with the rail bolts. The match between straight handrail and fittings is never perfect—some fairing is always needed. At left, a 2-in. dia. sanding disc attached to an electric drill is used to smooth the joint. In the photo above, the newel posts have been installed temporarily in order to calculate the cuts for attaching the easing between the gooseneck and straight rail. A section of straight rail has been attached to the easing and lined up with the vertical leg of the gooseneck to help position the easing correctly.

instead I use hex nuts and a twelve-point box wrench, as in the top two photos above left. This pulls them together carefully and firmly. But I don't glue them yet.

Goosenecks—Goosenecks form the transition from the incline of the stair back to level, either at a landing or at the top of the stairs. They come in two sizes, one-riser or two-riser, the difference being the length of their vertical leg. The size you need usually depends on whether the handrail simply levels off above the gooseneck or turns and continues to rise.

Sometimes two-riser goosenecks come disassembled so you can cut the vertical leg to whatever length you want. Single-rise goose-

necks never seem to be long enough, so I always use a two-riser gooseneck, just to be safe.

Instead of landings, some stairs have winders—treads that are wider at one end than at the other, used to effect a turn in a stair while continuing to climb. Winders are awkward and dangerous, so I try to discourage clients from using them. But if you do build them, be aware that they usually involve three rises at one point. You may have to add more straight rail to the vertical leg of the gooseneck to get enough height (photos p. 26).

The length of the vertical leg of the gooseneck depends on the position and height of the newel post. The tables supplied with the gooseneck fittings tell at what length to cut the verti-

cal leg, but again, that measurement is correct only if the newel-post locations and handrail heights are the same as in the diagrams, so check them carefully.

If the installation lays out the way the instructions indicate, the leg can be cut to the assigned length, the up easing joined to the leg with a rail bolt, and the curve where the up easing joins the straight rail cut off using the pitch block.

Lay the assembled gooseneck on the bench top with the back of the leg flat on the surface, set the pitch block on its rise edge and slide it against the underside of the up easing to mark the intersection point. Then turn the pitch block over on the run side. Scribe the angle through the first mark on the up easing, and cut it there;

Photos top and middle left: Rob Gruyé; Other photos: Sebastian Eggert

then attach the easing (without glue) to the gooseneck with a rail bolt.

You have to lay out everything in place again to determine where to cut the straight rail and attach the gooseneck fitting. Position the starting fitting over the newel-post location at the bottom of the stairs, with the straight rail attached and clamped in place on the treads.

Next you have to calculate the correct height to block up the gooseneck. If the height of the handrail coming up the stairs is to be 30 in., then the underside of the handrail will be 27¼ in. from the tread nosing (drawing, below). And if the height of the level handrail along the landing is to be 36 in., then the top of the newel post will be at 33½ in. The difference between these two measurements will be the distance from the underside of the gooseneck cap to the finished floor (33½ − 27¼ = 6¼).

Temporarily block up the gooseneck fitting this distance above the newel-post location. Now mark the point of intersection between the up easing you just cut off and the straight handrail coming up the stairs. Cut the straight rail, drill and attach the rail bolt, and assemble the gooseneck to the railing without glue. Now you can check all the fittings for alignment.

When the layout isn't exactly like the manufacturer's diagram, I stick with my empirical methods and lay everything on the stairs again, just as above, except that the easing isn't attached to anything. So now the straight rail is coming up the stairs and the vertical leg of the gooseneck is standing plumb at the top. The easing is simply an arc of a circle that's tangent to both of these rails.

Sometimes I can just hold the easing in place against the sides of the other pieces and calculate where to make the cuts by eye. But most of the time I attach a short section of straight rail temporarily to the easing and line this up with either of the other rails to find where to make the cuts (photo facing page, top right).

Newel-post installation—Once the handrail has been assembled, the newel posts can be cut to length and installed. With the assembled handrail lying on the stairs,

measure the distance between the bottom of the fitting and the stair tread below it. This measurement plus the height to the bottom of the handrail will be the height of the newel posts. Remember to add any additional length needed below the tread or landing to install the post.

The bases of the newel posts almost always have to be notched around the first step, into the corner at a landing, etc., so the alignment with the balusters is correct. Often the walls aren't plumb, so check them and lay out the notches accordingly.

I use 5/16-in. hex-head lag screws long enough to reach the rough carriage or other framing members and pull the posts securely to the wall. With three lags per newel in different directions, they should never come loose. Plug the countersunk holes with the same 1-in. plugs used with the rail bolts and line up the grain so they disappear. At times I've had to pull the flat bottom of the newel post directly to the subfloor with a rail bolt, and count on the adjacent sections of rail and newel posts to steady it.

The newel posts manufactured for use with starting steps have a long pin or dowel turned on the bottom. Usually you can attach the newel to the step first, and then install both as a unit to the base of the stairs. Start by drilling a hole the size of the pin through the tread, then through the starting step's horizontal core. Secure the newel with a lag screw and washer through the bottom of the starting step into the newel pin.

Sometimes I fasten the starting step and tread temporarily in position, check to make sure that the newel post is plumb, and glue it into place. When the glue has cured, I remove the newel post

and step in one piece to screw them together under the framework; then I return the starting step and newel-post assembly to its location for final installation.

I've done some jobs where the pin-bottom newel post wasn't long enough. I had to buy a regular square-bottom newel and cut a tenon on the bottom. At every step, double-check everything and proceed with caution, especially if you're in the land beyond the diagrams.

When all the newel posts are securely fastened, all mitered skirt boards, risers and treads can be installed. As you fasten the return nosing to the tread, try to avoid putting the screws or nails where the tread has to be drilled for the balusters. I seem to ruin a drill bit on every job by running into my own screws.

Handrail assembly—With the newels in place, check the fit of the handrail assembly. You may have to rasp the pins on top of the newels so the handrail fittings will slide onto them without pounding. Check that the posts aren't pushed out of plumb by a rail section that's too long, and that the fittings are sitting level on the newels.

When I'm satisfied that the handrail is accu-

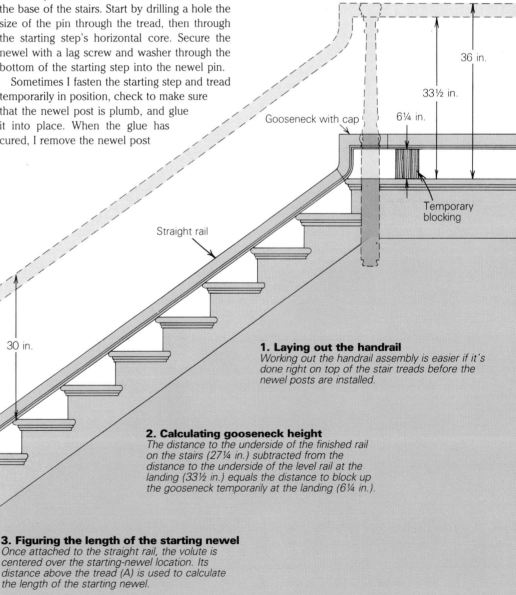

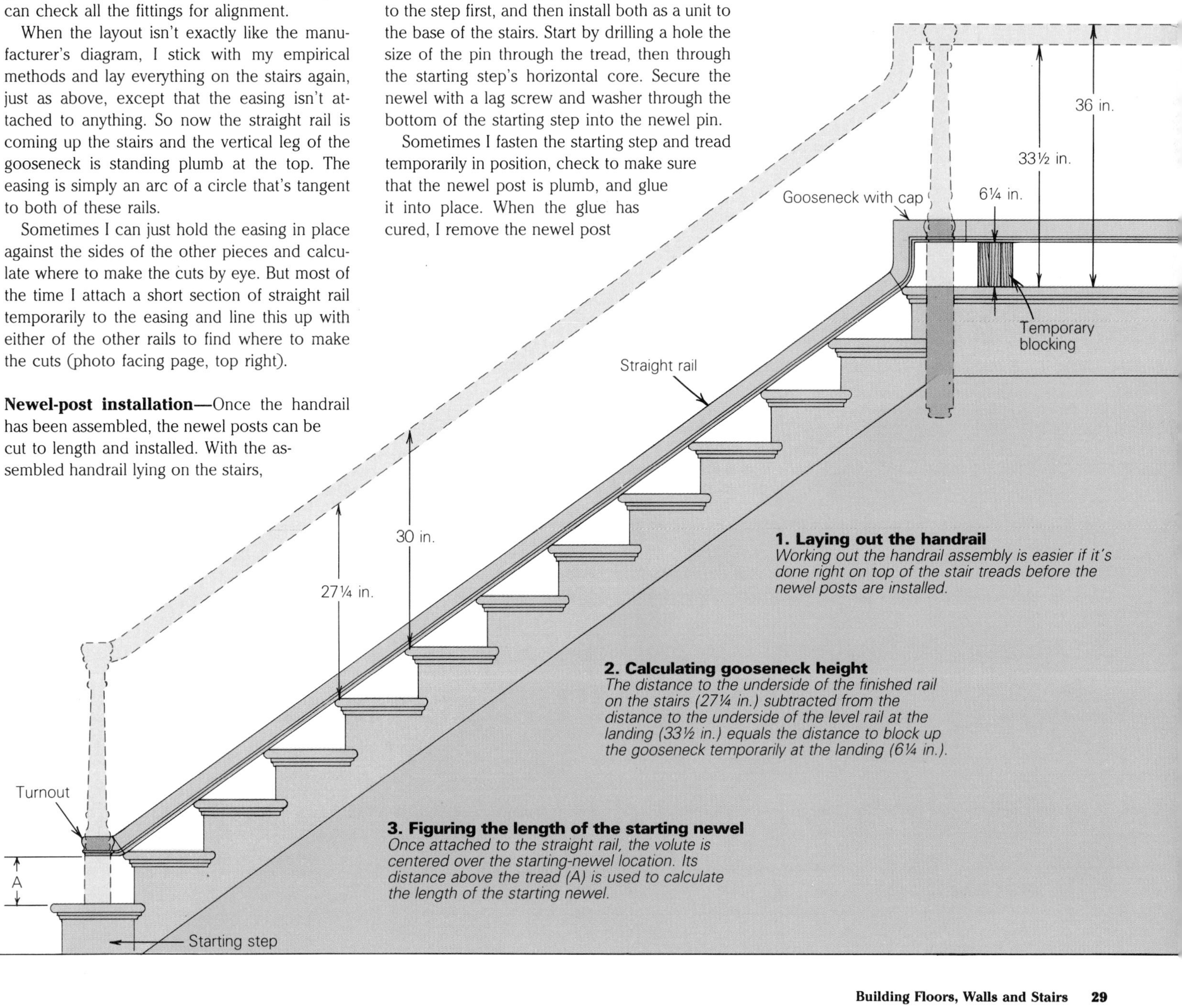

Gooseneck with cap

Straight rail

36 in.

33½ in.

6¼ in.

Temporary blocking

30 in.

27¼ in.

Turnout

A

Starting step

1. Laying out the handrail
Working out the handrail assembly is easier if it's done right on top of the stair treads before the newel posts are installed.

2. Calculating gooseneck height
The distance to the underside of the finished rail on the stairs (27¼ in.) subtracted from the distance to the underside of the level rail at the landing (33½ in.) equals the distance to block up the gooseneck temporarily at the landing (6¼ in.).

3. Figuring the length of the starting newel
Once attached to the straight rail, the volute is centered over the starting-newel location. Its distance above the tread (A) is used to calculate the length of the starting newel.

rately assembled, I pencil index marks across all the joints before I take it apart. These make it easier to line up the fittings when they're being glued. Where handrails level off, change direction and drop, the alignments are critical. A slight twist can throw off everything. If you're not sure that it will be a perfect fit when put in position, leave the joint dry and glue it together after all the balusters are in place.

If the handrail is fairly simple (one or two fittings), you can assemble the whole thing, plug the holes for the rail bolts, and fair and sand the joints before installing it permanently. If there are several changes in direction, you may have to assemble and install the handrail in sections. Assemble as much as possible beforehand, and plan the sequence of the final assembly to avoid problems tightening and sanding joints later. For instance, a joint at the vertical leg of a gooseneck and an up easing may be impossible to tighten in place, since the newel post will be in the way. Before the final assembly, I coat both sides of the joint with glue, then wipe off the excess with a damp sponge as soon as the parts are pulled together.

A heavy-tooth half-round file is useful for fairing the handrail joints, but I've also been using a small disc-sanding attachment on my drill to get into difficult places (photo p. 28, bottom left). The discs I use are 80-grit and 2 in. in diameter. I've heard that a Dremel tool with a selection of small burrs also works well for this, though it might be tough to fair out the joints smoothly with such a small tool. I have used a belt sander, but usually just on the underside of the railing to smooth off plugs.

After fairing the joints, I finish the sanding by hand with 100-grit and then 120-grit aluminum-oxide production sandpaper until all of the scratches are gone. The stair-part manufacturers say that the newel posts and balusters are ready to finish, but don't believe it. Go over them all just to make sure all the scratches are gone. I spin the balusters and newels on my lathe and sand them all with 120-grit paper for a really smooth finish. Don't use steel wool or sandpaper with a black grit on oak: it gets into the open grain and discolors it badly.

Baluster layout—Now that the newel posts are in place and the handrail is assembled, set the handrail up to mark the holes for the balusters. Traditionally, there are two balusters per tread, with the downstair face of the first baluster on each step in line with the face of the riser below it. The on-center spacing then should be half the run of the stair. I mark these locations on the treads, and plumb up to the handrail with a level or plumb bob, marking the underside for drilling the baluster holes. This layout can also be done when assembling the handrail on the treads, before the newel posts go in, simply by marking the baluster locations on the treads and then transferring the marks to the handrail with a combination square.

Some manufacturers suggest using a pitch block with a hole drilled in it as a jig to drill the holes in the handrail. But I just leave the handrail in place, line up the back of the drill motor with the mark on the tread below and drill carefully 1 in. into the handrail.

Sometimes a baluster layout will fall where a joint has to be made in the handrail. In this case, leave the baluster out until the joint is glued, plugged and allowed to cure. Then you can drill the hole for the top of the baluster. Cut the baluster just long enough to slide up into the top hole and drop into the hole in the tread. Glue and nail the baluster securely.

Always check the handrail for crowns or dips before cutting the balusters. I've had to pull crowns down by first toenailing a few key balusters and then pulling the rail down and toenailing the top of the baluster into the handrail. I measure the overall length of each baluster, subtract a whisker or two for glue, and cut each baluster to fit in its place. Be sure that the bottom pin is short enough for the depth of the hole in the tread and cut it off if necessary. The balusters I use come with a tip diameter of ⅝ in. for the top 3 in., so they usually slide into the handrail without trouble.

Final installation—With the balusters all cut, the real fun is about to begin. I apply a good grade of construction adhesive like Max-Bond or Liquid Nails to all the holes in the handrail and treads, making certain that there's good contact all around. These glues seem to allow a bit of flexibility without compromising strength (unlike yellow glues, which tend to be brittle and don't fill voids as well). I also smear some glue on the pegs of the newel posts and balusters. Too much in the holes may prevent the balusters from sliding all the way in.

Set all the balusters quickly into the tread holes, spin them around once to make good glue contact, and then, starting at the bottom, begin setting the handrail in place. Trying to get all those balusters to line up and slide into place is like trying to get a roomful of children to sit still and pay attention. The ideal helper would be an octopus. It helps to slide each baluster up into its handrail hole, but without pulling its bottom pin out of the hole in the tread.

When all the balusters and newel-post pins are in their holes, gently but firmly tap the handrail down with a rubber mallet (photo below left). The balusters may jump a bit, so you have to check that their bases are tight to the treads and that they are lined up parallel to each other. Some pins are turned a bit off center, and those balusters may need to be rotated 90° or 180° to line up with the others. Check their alignment with a straightedge. Eyeball down the length of the rail for crowns and adjust with the mallet. Make sure the fittings are all the way down on the newel posts.

Sometimes I'll drift a long screw from the side of the fitting into the newel post to pull them together. Countersinking one through the top is easier, but if you do this be sure to find a plug that matches the wood grain around the hole and line it up carefully. Staircases are great rainy-day jungle gyms, so in houses with kids, I'll toenail all the balusters, top and bottom, just to be on the safe side.

After all the rails are on and the balusters are in, check for scratch marks from assembly and sanding. I use penetrating oil stains and clear finishes, applying several coats to bring up a smooth shine. Some parts like balusters can be prefinished, but wherever there is a nail hole to be filled or a plug to be sanded it can botch a nice finish, so it's better to finish the whole staircase after it's installed.

Take your time and do the best job you can. A beautifully crafted staircase is the centerpiece of a house, a delight to the eye and to the touch, and an example of the finest skill a carpenter has to offer. ☐

Lining up the balusters and worrying them into place for the final installation is a tricky job. It helps to have an extra pair of hands. Tapping with a rubber mallet ensures that the handrail and fittings are all the way down on the balusters and newels.

Sebastian Eggert builds staircases and mantels in Port Townsend, Wash.

Octagonal Spiral Stairs

A complicated stairway built and installed in separate halves

by Tom Dahlke

The opening for this stairway was an elongated octagon in the center of an octagonal house. The floor on the main level was laid in wedges to emphasize the unusual shape, and the owner wanted an octagonal spiral staircase to carry out the idea. The completed stair makes a full 360° spiral. Its unsupported stringer describes an eight-sided figure that's echoed by the rail.

The house Gary Therrien built for himself is octagonal, with a truss-roof system that required no load-bearing interior walls. The floor of the large, open main level emphasizes the shape of the house: oak boards laid to form wedge-shaped sections that were edged with mahogany strips running to the house's corners. In the center of the floor is yet another octagon: the opening for the staircase that leads down to a lower-level bedroom, the entry, workshop and playroom. Therrien wanted a spiral stairway, but the curves of the standard round or oval spiral wouldn't work with the angles and straight lines of the rest of the house. He felt, and I agreed, that it would be more appropriate to build a stair whose outside edge followed the shape of the opening in the floor. The fact that this octagon didn't have equal sides was something I didn't want to think about yet.

The basic spiral staircase isn't all that hard to lay out or build, but I was dreading the thought of putting together the unsupported octagonal stringer for this one. I also realized early on that the walls around the opening on the bottom floor would make it impossible to stuff a fully assembled unit in place, I would have to build the stairway in two vertical halves and install them one at a time.

The opening was elongated—6 ft. 6 in. wide by 7 ft. 6 in. long. Each end had three 31-in. sides joined by two 43-in. sides. The stair treads had to spiral 360° over a total drop of 9 ft. 3 in. I decided on 14 treads with a rise of 7⅜ in., and began by drawing up a plan view

Assembling the post. The two halves of the post were coopered up separately. Some of the blocking inside is reinforcement, right, and some will support the octagonal blocks that the treads will be bolted to. The blocks also help align the halves when they are assembled. Mortises, each one different, have already been cut for the treads. Above, the post is strapped together dry, before the treads are installed.

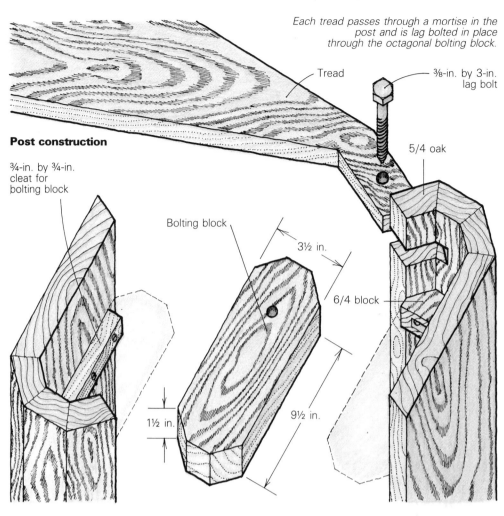

Each tread passes through a mortise in the post and is lag bolted in place through the octagonal bolting block.

Tread

⅜-in. by 3-in. lag bolt

5/4 oak

Post construction

¾-in. by ¾-in. cleat for bolting block

Bolting block

3½ in.

6/4 block

1½ in.

9½ in.

and two elevations that showed where each tread would join the post and at what angle. The treads were laid out at 22½° intervals.

I set up my table saw (which has a mortising attachment), my jointer-planer, and my bandsaw in the living room, and did all the work on site.

The hollow post—I planned to rabbet the treads into the outside stringer. Since most of the stringer would be unsupported by anything else, the treads had to be locked immovably to the central post. Because of this, and because of the space problem, I decided to cooper up a hollow post. This would let me fasten the treads very sturdily on the inside. It also made it fairly simple to assemble the stairway. I would glue up two vertical halves

of the post, and bolt the treads to 1½-in. blocks shaped to fit inside the hollow space (drawing, above). These bolting blocks would rest on cleats screwed into the post. Other bracing would serve as guides when I glued and screwed the post halves together.

I used eight boards 9¼ ft. long and 1¼ in. thick to make the hollow octagonal post. To get the correct sectional dimensions, I ripped two of them to a width of 8½ in. and six to a width of 2½ in. Next, I tilted the blade on my table saw to an angle of 22½° and beveled the two edges of each piece. Then I put them together dry with strap clamps while I marked the post for the mortises. I had to scribe each mortise on both the outside and inside of the post to determine the angles at which the treads were coming in. To do this, I marked

their locations in plan view on one end of the post, and measured down the inside and outside surfaces. I also drilled holes every 15 in. along the edges of the two halves of the post. These were for the countersunk screws that would hold the pole together after assembly. Then I took the strap clamps off, and cut the mortises on each half with the bandsaw where possible, or with handsaws and chisels. No two mortises were alike, and I had to cut them precisely so the treads would mate with the stringer. When all the mortises were cut, I strapped the halves together dry again to have a look (photo above left).

Treads and stringer—No two treads are alike, either. Their outlines depend on the angles at which they meet the post, and the

Photos p. 31: Donna Coveney; Illustration: Vince Babak

The treads were installed while the two halves of the post lay across sawhorses, top. The stringer sections, which had previously been routed out, were temporarily attached to the treads, and then carefully trial-fitted, above, to achieve a smooth joint between sections.

shapes imposed on their outer ends by the rising and angling stringer. I cut the treads long enough to fit into ½-in. deep mortises in the stringer, then made sure each one fit the post. Finally, I rounded their edges with a router and finish-sanded them with 220-grit paper. With the post halves resting on sawhorses in the cellar, I fit the treads into their respective mortises and bolted them to the interior blocks with 3-in. lag bolts, as shown in the photo at left.

Treads in place, I began cutting the stringer to fit their ends, and from here on, my drawings were worthless. Each joint required a compound-angle cut. I knew that all the miters had to be 22½° (actually 67½°, but for setting the blade angle on a table saw, whose protractor reads from 0° to 45°, you have to use the complement of 67½°, which is 22½°) for the stairway to make 135° turns, but the vertical angles varied because the rise of the segments varied. (The segments of the octagon were not equal, and this was the only way to be sure that the front and back edges of the treads would remain within the width of the stringer.) I'm sure there is a way to figure this stringer out on paper but I could only get part of it, and wound up trial-fitting instead.

I used three 12-ft. oak 2x12s for the stringers, with the best one in the middle where it is most visible. I rough-cut the segments to length one after the other down a board so the grain would be continuous on both sides of a joint. Then, with the post halves bristling with treads still on the sawhorses, I used my drawings and some simple calculations to lay out the mortises on the stringers that would house the wide ends of the treads. I routed out the mortises and set the rough-cut stringers in position on the treads. I predrilled each segment so it could be bolted to the treads with 3½-in. lag bolts countersunk and plugged with 1-in. oak plugs.

From here on, I could concentrate on the joints between stringer segments. I started at the bottom and worked up, cutting and recutting each segment until all the joints were tight (photo bottom left). I predrilled them for six 2-in. screws, to be countersunk and plugged with ½-in. oak plugs. Then I beveled the stringers' edges and used my table-saw mortising jig to drill dowel holes for the rail balusters. Finally, I sanded the sections with 220-grit paper.

Installation—On the day of the raising we had ten people on site, and we needed every one of them to muscle the two halves of the stairway into place. The stringer segments that could be installed later were left off, but the segments at the top had to be installed before the stair could be positioned. At the third and fourth steps from the top, the stringer passes within ¼ in. of a wall, and after we'd maneuvered the first half in where it belonged, we lag-bolted through the stringer segment into the studs. Two 1-in. dia. oak pins through the bottom octagonal bolting block hold the post in place on the floor.

With one half secured, we slid the other

into place to check the fit one last time. It mated perfectly, so we glued, clamped and screwed the halves together.

After the remaining stringer segments were glued, screwed and bolted in place, I plugged all holes with oak plugs and sanded them smooth. The bottom section was drilled to sit on a 1-in. dia. oak peg set in the flooring.

The top of the post is held in place by a landing support cut from the stringer stock. A one-of-a-kind joint was cut so the 2x12 oak would join the post flush. To continue the effect of the spiral, we built the landing in two tread-size sections to give the illusion of two more steps.

Bannister and balustrade—The owner wanted the bannister to look like a continuous, floating ribbon, beginning over the post, going around the opening, then down and around the stair. I bought five rough-cut 14-ft. full 2x6 mahogany boards, then resawed them down to 1¾ in. by 5 in. I took the lengths and angles directly from the stringer, and sawed each segment so the grain would carry around the turns. I beveled the top edge of each section to get three surfaces of equal width, and routed a ¼-in. bevel on the bottom edges. All of the segments were predrilled to accept the dowels for the balusters. The joints were glued and screwed where they were easily accessible, and splined where it was harder to get at them. The connection between the horizontal rail at the top of the stairway and the first angled rail was made with dowels for the sake of simplicity.

The balusters are 1¼ in. square, and they are about 6 in. o.c., but because the sections of rail are different lengths and I wanted the balusters to be evenly spaced along each section, I had to vary the spacing a bit. I drilled ½-in. dia. holes 2 in. deep in both ends of each baluster to accept ½-in. dowels. Normally, I would mortise a baluster in, but the stairway's very shape helped make the rail so sturdy that the doweling, which is much simpler, was strong enough.

The balusters couldn't be cut to a single length to fit between the bannister and the stringer. I had to custom-cut each one to length, in the process making sure that I cut the correct angle on each one's top and bottom. To get the clean ⅛-in. bevels on all four edges, I hand-planed each baluster.

We had trouble deciding how to resolve the top end of the rail. The post ends at floor level with an oak and walnut burl cap sitting on top. At first we thought the rail would look good if it appeared to emerge out of the post. Well, once done, it didn't, so I brought out a chainsaw and did some remodeling. The result was to have a short 6-in. piece dip down at the same angle as the piece opposite where it starts to go down the stair. A much better solution. At the base of the stair, the rail simply ends with the last tread. □

Tom Dahlke is currently building timber-frame houses in Canton, Conn. Photos by the author, except where noted.

Three Custom Stairbuilders

The secret lore of stairbuilding still lives,
but initiation can be tough

by Kevin Ireton

The best book currently available on stairbuilding was written over 80 years ago. It's William and Alexander Mowat's *A Treatise on Stairbuilding and Handrailing*, and it has been out of print until just recently. But three years ago bookseller Richard Sorsky tracked down an original copy of the Mowats' book at a London book fair and reprinted it. So far, Sorsky's company, Linden Publishing (3845 N. Blackstone, Fresno, Calif. 93726; 800-345-4447) has sold more than 6,000 copies.

In his introduction to the book, Sorsky calls the late 19th century the golden period in the art of stairbuilding. Since that time, the steady encroachment of technology, and a work ethic that values white-collar over blue-collar jobs, has meant the decline of stairbuilding and the near-extinction of stairbuilders. As manufacturers began to mass-produce stair parts—newels, balusters, easings, goosenecks and rails—staircase designs were simplified to accommodate these less expensive parts.

This decline accelerated after World War II, when the demand for housing encouraged quicker and cheaper construction. The proliferation of single-story houses nearly eliminated the need for stairs altogether. The collective literature on the subject went out of print, and many of the master stairbuilders died without passing on their knowledge.

Over the past 15 years there has been a resurgence of interest in stairs and stairbuilding. Partly a consequence of the restoration movement—where 100 or 200-year-old staircases need rebuilding—the burgeoning interest in stairs is also due in part to the baby boomers. With double incomes and smaller families, they have more money to spend on houses, and custom staircases, long associated with elegant homes, are something they want.

My own interest in stairs began while I was working as a carpenter. With the possible exception of boatbuilding, stairbuilding is the most exacting discipline in large-scale woodworking. It's certainly the most challenging work that a residential carpenter ever faces. I took on every stair job I could, whether it was rebuilding a Victorian balustrade or simply cutting rough stringers for a set of basement steps.

About that time, I also began to hear of custom stairbuilders whose skills were so rare and highly developed that others sought them out from all over the country. Their domain, for the most part, was grand curved stairways. They did 90% of the work in the sanctity of their workshops, then shipped the staircases to their destinations and installed them.

Since I had never met one of these stairbuilders and didn't even know for certain that they existed, their work assumed a mythical stature in my mind. Now that it is my business to write about houses rather than build them, I decided to track down some stairbuilders to learn about their lives and their work. And given the disruption of the craft, I especially wanted to know how they learned their trade.

Having built a scale model of the stair in the photo above, Mulder knew it would fit through the front door. But he hadn't expected the decorative trusswork overhead. The stair fit, but it took 20 men to screw it through the opening. Like many contemporary stairbuilders, Mulder laminates his handrails. Unlike most of them, he glues a cap on top to hide the laminations. In the photo at right, he's using a router to trim the cap flush with the sides of the rail.

From cabinetmaker to stairbuilder—When David Mulder decided to go into business for himself as a stairbuilder, he designed a stairway to showcase his skills. Then he hired an architect to design a house around the stairway. Mulder spent the next two years building the house and stair, working by himself twelve and fourteen hours a day, six and seven days a week. Bringing a cabinetmaker's sensibility to carpentry, Mulder got fanatical at times. After sanding his new oak floors, for instance, he hand scraped behind the edger to remove the nearly imperceptible circular scratches that most people ignore.

The stairway itself, shown in the photo on the facing page, is one of the nicest I've ever seen, chiefly because of its scale. Most curved stairways are grand things, dominating entry halls with 30-ft. ceilings. But Mulder's is an intimate stair. You wouldn't feel underdressed coming down these steps in the morning wearing your bathrobe. Still it's the focal point of the house, and a dramatic presence.

This circular oak staircase rises from the finished basement to the second floor of Mulder's own house in Battle Creek, Mich. At the first and second-floor landings, a 4-ft. wide swath of strip-oak flooring sweeps, like sun rays, around the curve of the stairwell, leading from one flight to the next.

The staircase isn't a masterpiece, but is certainly the work of a proficient journeyman. And Mulder himself, an extrememly modest man, readily admits this. When I approached the stair for the first time, he quickly pointed out the few flaws before I could notice them—one of the volutes, for instance, is slightly misaligned with the bullnosed starting step. Depleted at the end of the project and anxious to get his family moved in, Mulder had rushed the handrail installation. He was careful to add that he tolerated such mistakes only because it was his own stair, not one for a customer.

After he finished junior college, Mulder worked with a local stairbuilder for several years—a self-taught man who, by his own admission, was "no cabinetmaker." Later Mulder worked in a custom cabinet shop for six years, and that's where he learned about wood, joinery and tools. There he developed the skills that enable him to build stairs.

A five-minute drive from his house, Mulder's shop is a large pole barn with a 16-ft. ceiling and enough room to build two staircases. The shops sits well off the road, in the middle of a 15-acre wooded lot where Mulder hopes one day to build another house around an even more impressive stairway.

After he finished building his present home, Mulder had the staircase photographed for a

From *Fine Homebuilding* magazine (December 1987) 43:34-39

Facing page: Michigan stairbuilder David Mulder dreamed up this staircase, then hired an architect to design a house around it. He spent the next two years building both. But since it was his own house as well as the basis for a successful advertising campaign, the effort was worth it.

brochure advertising his services. From somewhere he inveigled a list of practicing architects around the country and started mailing them his brochure. He's been busy building stairs ever since. But his isn't a high-volume shop; Mulder hopes to do only six staircases this year.

Mulder works from blueprints and from other people's measurements. It's a risky thing to do, but he considers it a pragmatic concession. "Before I start the stairway," he says, "I make sure the carpenter and everybody else understands how critical the measurements are. I want them to know exactly how I'm building it." After receiving the blueprints, Mulder lays out the staircase full scale on the floor of his shop. Then he does his own ¼-in. scale drawing of the stair and sends it back to the carpenter who took the original measurements. "He double-checks me and I double-check him. It works out okay."

The crux of custom stairwork is making rails. Traditionally, curved handrails were sawn and carved in sections from large blocks of solid wood, and then joined with dowels and rail bolts. But laying out these sections, carving their molded profiles, joining them squarely and achieving a fair curve that exactly follows the line of the stringer requires tangential geometry and full-scale developmental drawings.

The formal system for doing this work was developed early in the 19th century by an Englishman named Peter Nicholson. Most of the books on stairbuilding, including the Mowats', focus on making rails and are based on Nicholson's system. But learning to make handrails from these volumes is akin to learning brain surgery from a book.

I asked Mulder if he was familiar with the Mowats' book, and he said that he had a copy but hadn't read much of it. He doesn't have any patience for the cryptic language and arcane methods, not when there's an easier way.

Like many of the stairbuilders working today, Mulder laminates his handrails from thin strips of wood wrapped around the same form used to make the stair stringers. One objection to this method is that all of the plies are visible and the rail no longer has the look of solid wood. Mulder conceals the plies by laminating a thin cap on top of his handrails.

After removing the laminated rail from the forms, he squares it up with a hand power plane and smoothes it with a scraper. To cut the profile, he uses a huge router (Bosch 3¼ hp, model 90300), fitted out with custom-made bits that look like shaper knives (bottom photo, previous page). It is an intense process.

"I can't have the dog chasing the cat, or my son chasing the dog," he says. "I have to be alone, with no interruptions. I go by hand, by eye and by ear. I listen to the wood. If I hear a change in the sound, it may mean the grain has changed directions, and I'll have to stop and rout the other way."

But the router, riding along the squared top and bottom of the handrail, can shape only the sides. To mold the camel's hump that runs down the top of his handrails, Mulder uses an electric grinder, a scraper, files and rasps. He uses no template to check his work, but goes strictly by eye and by feel. "If I checked my work with a

A 100-year-old English text on stairbuilding, a brass quirk router made by his father and a home-made wooden plane are among the tools Timothy Johnson uses to do work like this walnut volute.

template, it might be wrong—one section slightly different from another. But if it's pleasing to the eye and to the touch, it can't be wrong."

Mulder's forsaking traditional methods in favor of his own doesn't necessarily compromise the quality of his work. Unburdened by the weight of tradition, he simply asks what the best way is to do each given task. His method of attaching return nosings is a good example. Traditionally, return nosings are merely nailed on the end of the tread. Mulder, however, glues them on with a spline in the joint, which is stronger and leaves no nail holes.

The lack of nail holes is a point of pride with Mulder. Except for the cove mold under the treads, which he puts on with brads, Mulder leaves no sign of fasteners on a stair. All screws and nails are either applied from underneath or else hidden behind something else.

He also has a cabinetmaker's eye and concern for grain orientation. His decorative brackets, for instance, are cut from a single board and installed in sequence so that the grain flows from one to the next.

Except for the balustrade, Mulder usually assembles the entire staircase in his shop. With the help of friends, he loads it on a flatbed trailer (along with the handrail, balusters and other miscellaneous parts), hitches the trailer to his 1-ton Suburban and hits the road.

Mulder hauled the staircase shown in the top photo on the previous page to Wichita, Kan. When he pulled into the driveway, the first thing that he noticed was the front door of the house. He knew the stair would fit through the rough opening because he'd built a scale model and tested it. But no one had ever mentioned the decorative trusswork on the overhanging gable

above the door. Oversights like this keep stairbuilders awake at night.

As it turned out, the staircase did fit. It took 20 men to corkscrew it through the door, but they made it. Mulder installed the carcase in a day, carefully explained the balustrade installation to the lead carpenter and headed home without ever seeing the completed stair.

When prevailed upon, though, Mulder will install the balustrade himself, but he'd rather not. In part it's because he dislikes being away from home and family for too long, in part because he's anxious to get started on the next stair.

A fundamentalist—Some years ago, Timothy Johnson set up a booth at a local home show to get work for his carpentry and cabinetmaking business. It was the First (and last) Annual Franklin, N. C., Home Show, and Johnson admits that his being there was a desperate act.

He built part of a staircase—three steps, a newel post, a handrail and a few balusters—and set it up in his booth to display his skills. People responded, especially older people, running their hands over the railing, smiling, saying they hadn't seen anything like this in 50 years, or that they'd grown up with a staircase just like this. He got four stair jobs as a result. Since he hadn't had much luck peddling his furniture, and since he didn't want to make jewelry boxes or kitchen cabinets, Johnson began to wonder if there wasn't a market for custom stairs.

While working on those four initial stair jobs, Johnson remembered reading an article called "Building Stairs" in the September 1981 issue of *Fine Woodworking* magazine, whose back issues he saved. He dug out and studied the article, which described the techniques of the late Harry Waldemar, a master stairbuilder from New York. With this, Johnson began a research effort that continues to this day.

The article taught Johnson a lot. It also generated an editor's note in a subsequent issue of the magazine, listing several old books on stairbuilding, all out of print. After six months of searching, a local bookdealer found one of these— *Stairbuilding* by Gilbert Townsend (1946).

Townsend's book got Johnson started on curved work and introduced him to the traditional method of making rails. But the information was spotty and obscure. So Johnson had his bookdealer track down other stair books, and eventually he acquired a copy of Fred Hodgson's *Common-Sense Stairbuilding and Handrailing* (1903). The title proved ironic, as Johnson found the book more frustrating than helpful.

When Richard Sorsky reprinted the Mowats' book, Johnson was among his first customers. The arrival of the book in Johnson's shop was a milestone. It was the most thorough and lucid explanation of the subject he had ever seen. As Johnson puts it, "It was as though you were interested in light bulbs and suddenly had Thomas Edison for a teacher."

"The works required an evangelical dedication from the reader," Sorsky writes in the introduction to the Mowats' book, referring to the early texts on stairbuilding. Johnson was such a reader. A former seminary student who had majored in Greek, he had a reverence for tradition and patience for translation that prepared him for the long hours he spent wading through the arcana of the ancient texts.

Still, he was working alone, and progress was slow. But he was relentless. When he read that wreathes and volutes were cut out with a bowsaw, Johnson hadn't ever used a bowsaw. But he found the plans for one, made his own and now cuts out his wreathes and volutes with it.

Similarly, he read about a quirk router, used for running rabbets on curved stock. No one he asked knew what a quirk router was, but eventually he found a picture of one. And that was all he needed. Johnson enlisted the help of his father, a retired machinist, and made one out of brass. It looks like a spokeshave, but works like a router plane (photo facing page). Most of the curved rail parts he does entirely by hand, using a compass plane, various homemade planes, spokeshaves, riffling files, and of course, his quirk router.

But while there's a legitimate place in stairwork for hand tools and Johnson clearly enjoys using them, he isn't a hand-tool purist. He has the requisite machines and uses them to good effect. He grinds his own knives and runs all his straight rail on the shaper. And he's got a sophisticated duplicating lathe, similar to a Hapfo, that he and his father built from a machinist's lathe. It turns out finished balusters in a single pass, and saves a lot of drudgery when the time comes to turn a hundred or more balusters.

Johnson likes the practicality of stairs. They're not just curios or art pieces, like some of his cabinets and furniture; they're a necessary part of a house. And in his pursuit of traditional methods, he's found some reassuring good sense. Balusters, for instance, were originally

Johnson built and installed this suspended stair in a home on the north side of Atlanta, Ga. With the help of a small drawing in an old stair book, he devised a method of cross bracing the underside so that there is no spring when you walk on the stair.

dovetailed into the stair treads. Johnson is the only stairbuilder I met who does this (others simply turn a pin on the bottom of the baluster and glue it into a hole drilled in the tread).

Johnson denies that there's anything impractical in this. He argues that dovetailed balusters make installation a more relaxed process because you don't have to line up all the balusters at once. You can put the handrail in place and install the balusters one at a time. Also, by removing the return nosing, you can easily replace a single baluster. In fact the return nosing on the end of stair treads originated as a way to reach and cover baluster dovetails.

Dovetailed balusters also make a stronger balustrade, which is especially critical if there are no newel posts. Johnson told me, "You can't rely long term—60 or 80 years—on a baluster that's been pinned and glued . . . Let me put it this way, I won't ask someone to do that."

Johnson cuts the dovetails with a jig on his bandsaw. The housings in the tread he does with a backsaw and a chisel. Johnson's one concession to modern technology here is to run a drywall screw through the dovetail into the tread.

In the Mowats' book, Johnson read a reference to still another book, James Monkton's *National Stair Builder* (1872). When he finally found a copy, it cost him $125. But before Johnson finished reading the introduction, the book had paid for itself.

When I visited him last April, Johnson had just finished the installation of his most ambitious stair to date—a 28-ft. long suspended staircase, open on both sides, with a quarter-turn and flanked by open railings on the second floor (top photo, previous page). It commands the foyer of a new Greek Revival mansion on the north end of Atlanta, Ga.

Like most custom stairbuilders, Johnson designs the stairs that he builds. At this level of complexity many architects will gladly defer to an expert. So it was up to him to decide how to support the suspended staircase. Despite his fa-

ther's assurances that if the stringers were sound and well anchored top and bottom, everything would be fine, Johnson still worried. He had to *know* that it wouldn't sag in the middle, that it wouldn't even spring when walked on.

Here's where the Monkton book paid for itself. In the introduction, Johnson found a drawing about the size of a postage stamp and a short paragraph about "timbering the undersides of suspended stairs." Again, it was all he needed. By the time I visited, the underside of the stair had been plastered, so I didn't get to see the timbering that Johnson had done. But I can testify that the stair didn't move when I stopped in the middle and jumped up and down.

Handed down—Although my preference runs to small shops, I felt obligated to visit at least one larger stair company. I settled on Harmonson Stairs, a 26-person shop in Mt. Laurel, N. J., just outside of Philadelphia. Bart Withstandley, the owner, had told me over the phone that in addition to several hundred straight stairs last year, they built 130 circular staircases, and he predicted 200 for 1987. I arrived at Harmonson expecting a small factory. What I found instead was an enclave of traditional stairbuilding.

I saw young men, with less than two years experience, making rails the traditional way—cutting out sections on the bandsaw, cranking them into bench vises and shaping them with planes and spokeshaves (photos below). These were rails as complex in shape and profile as any I had ever seen. But at Harmonson, there was no awe or wonder in this. Certainly the work required a great deal of skill, which had taken time to acquire, but it involved no guesswork, no evangelical dedication to old books. Someone had simply shown these guys how to do it. That someone was Joe, Sr., the head layout man at Harmonson.

Joe started building stairs in 1941, and except for two years soldiering in World War II, has been building them ever since. On my first visit

to Harmonson, I didn't meet Joe because he'd taken the day off, which I later learned was no coincidence. He's publicity shy, for one thing. But he's also exceedingly purposeful, with no tolerance for the inept or the impractical, and he brooks no interruptions when he's working.

When I conspired to meet Joe on a subsequent visit, he ignored me at first, pretending to be busy with other things. A tall man, no stoop in his shoulders, there is a Gary Cooper quality about him. With steel-rimmed glasses and a full head of grey hair, I also saw why the men in the shop had said he "looks like knowledge . . . looks like the *Encyclopedia Britannica*." Eventually Joe poured a cup of coffee and headed for his office. Halfway there, he turned back to me. Looking like a dental patient about to suffer an extraction, he said, "Come on."

In 1951, the layout man where Joe was working died of a heart attack. Joe, who was in the field at that point, building and installing stairs—"working outside" he calls it—was brought in to be the new layout man. Since his predecessor had died suddenly, Joe had to learn largely on his own. His primary source of information, which he still uses today and refers to as "the bible," was Robert Riddell's *The New Elements of Hand Railing* (1900).

Joe came to Harmonson in 1974, when the company he'd been working for went out of business, and brought with him the technology to do curved work. Until then, Harmonson had done mostly staight stairs.

Today the company turns out five curved stairs a week. Withstandley thinks of Harmonson as a semi-custom shop. While they build totally custom stairs when called upon (photo facing page), most of their curved stairs are variants on a half-dozen or so models. The difference in cost—from a semi-custom model to totally custom—is anywhere from $4,500 to $40,000 and up.

Walking through the shop, I noticed a few factory-like compromises. For instance, the decorative brackets on production-model stairs are made of thin plywood and aren't mitered to the riser; they're merely applied to the outside of the stringer.

But such concessions to production were the exception. There was far more that impressed me, like the fact that Harmonson varnishes the underside of all stairs before shipping them.

As layout man, Joe is at the head of production. Below him the company is divided into stairbuilders, railmakers and installers. I was pleased to learn, after talking to these specialists, that they all respect each other. Even so, they expressed envy for each other's jobs. The railmakers wish they knew how to install stairs, the installers wish they knew how to make rails, and so on. This specialization was probably the worst symptom I saw of Harmonson's size and high-volume production. Just as I knew that Timothy Johnson would envy the Harmonson employees' relationship with Joe, I knew that they would envy his total involvement in the building of each staircase.

Given the general tendency to venerate the craftsmanship of bygone days, I was surprised to hear Joe say that the stairs they're building today are better than the ones he built 40 years

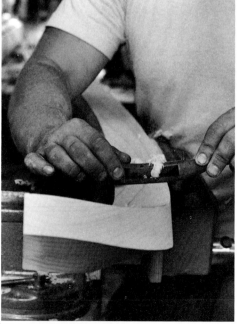

Under the tutelage of a 67-year-old master stairbuilder, young workers at Harmonson Stairs make handrails the traditional way, using a system developed in England 150 years ago.

The balusters on this Harmonson stair had to vary from two per tread to four per tread in order for the spacing to look right.

ago. Among other things, the joints are tighter. He thinks it's because more work is done in the shop now, where conditions are controlled. Joe has the carpenter's habit of pointing with his folding rule. And later, in the shop, he pulled one from his back pocket, opened it 2 ft. and pointed at the joint between the treads and stringer on a finished stair. "We never could have got it that tight 40 years ago."

The sophistication of the machines and their use, I'm sure, has also improved the quality. I've never seen tighter return nosings on treads than I saw on Harmonson's stairs. They cut them on the shaper, so instead of being a straight line the top of the miter is radiused, which adds an extra flourish. They have also worked out a way to mold most of their curved rail parts on the shaper. After being squared by hand at the benches, the pieces are fed into the shaper knives, riding across a fixture called a saddle. Instead of providing a table to support the work, the saddle provides a ridge, which allows curved pieces to pass over it.

Years ago, Joe had a young apprentice work-

ing under him. "He wanted to be a rail man," Joe said. "He was in a hurry." Joe explained to him that the curves had to be smooth, that he had to use his eye. After the apprentice had roughed out his first rail, Joe spotted a kink in it, but let the man finish it, laboriously molding the profiles. When the apprentice presented the rail to Joe for his approval, Joe said, "Good, now throw it the hell away and make a new one."

I wasn't surprised by the story. Five minutes after meeting Joe I could have guessed he was a hard-nosed taskmaster. But it did make me wonder about the men working under him now— did any of them have the patience to learn this difficult trade from such a curmudgeonly teacher? At least one of the men at Harmonson, I knew, had probably made a lifetime commitment to stairbuilding.

I asked Joe about his son, who at age 32 is now the head rail man at Harmonson. "I didn't think he was going to make it at first," Joe told me. But I knew Joe well enough by then to figure that this was his backhanded way of saying his son was doing all right. □

Custom stairbuilders?

There are no guilds or trade associations devoted to custom stairbuilders. Some of the larger stairbuilding companies belong to the Architectural Woodwork Institute (2310 South Walter Reed Dr., Arlington, Va. 22206). A few are listed in *Sweet's Catalog File* (McGraw-Hill, 1221 Ave. of the Americas, New York, N. Y. 10020) and a few in the *Old-House Journal Catalog* (The Old-House Journal Corp., 69A Seventh Ave., Brooklyn, N. Y. 11217). Some companies advertise in architecture magazines. And most stairbuilders are probably listed in the Yellow Pages wherever they work.

We know of no comprehensive list of custom stairbuilders around the country. In order to compile such a list, we would like to hear from people who build custom stairs. If this work makes up the majority of your business, send me a letter briefly describing your work, and perhaps enclose a photo or two.   —*K. I.*

Photo this page: Tom Crane © 1985

Plunge-Router Stairs

An adjustable mortising jig simplifies the construction of an open-riser stairway

by Bill Young

Leaving the risers out of a stairway can make a big difference in some houses. The open spaces between the treads allow for air circulation, and also a way for natural light to get into a sometimes claustrophobic corridor. Last summer, I built a pair of these riserless stairways for a three-story house, and it gave me two good problems to solve.

First, I wanted to mortise, or house, the treads into the stringer rather than notch them. I think housed treads look better, and this method also leaves the stringer intact, allowing for longer spans and a stairway with less bounce. The problem was how to cut the mortises quickly and accurately.

I knew all along that I would be making these cuts with my plunge router. This kind of router allows the bit to be lowered directly into the work, rather than let into it from the side. My first job was to figure out what kind of a jig would do the job precisely, yet still be easy to move along the stringer.

The second problem was to make the jig adjustable. Although the rises and runs for the two stairways were nearly equal, they had enough variation from floor to floor to require a different setup for each pair of stringers. My solution to both of these problems is the device I call the universal stair-stringer jig (photo facing page).

The jig consists of two basic parts: a rotating circular template and a cradle. A slot centered on the template's axis duplicates the cross section of the tread material, and guides the bit. Because I use a router bit that has a ball-bearing pilot mounted over the cutter—instead of under the cutter as is the usual case—I can size the cutout in the template to the exact profile of the stair tread. The cradle is a rectangular piece of plywood with a circular hole that is slightly larger than the template. Parallel fences on the bottom of the cradle position the jig on the stair stringer. So that it will be easy to move the jig along the stringer, I put the fences about 1/8 in. farther apart than the width of the stringer.

To use the jig, I lay the cradle over the stringer and I rotate the template until it aligns with the tread layout. This is where the universal part comes in. The jig can accommodate any tread layout, no matter what the rise or run of the stairs. When I'm satisfied with the alignment, I shim the fence against one side of the stringer (photo facing page). Then I use a bar clamp to lock the template inside the cradle, and to hold the jig in place as the stringer is routed.

Building the jig—I use 1/2-in. MDO plywood for my stair jigs. The letters stand for medium-density overlay. The overlay is a thin layer of plastic, which gives the plywood a smooth, hard finish. Around here a sheet of it costs less than $20. I've found MDO to be the perfect material for router jigs because it's strong and stable, it takes layout marks well and the router glides easily across its smooth finish.

I began building this jig by making the cradle from a 16-in. by 24-in. piece of MDO plywood (drawing, below left). I scribed a 13-in. dia. circle onto the rectangle, leaving three equal margins at one end of the rectangle. The wheel could be larger, but not much smaller. My treads were 11 in. wide, so a 13-in. template wheel gave me room for 1 in. of material at each end of the mortising slot. I drilled a 1/2-in. starter hole just touching the inside edge of the scribed circle, and I adjusted my electric jigsaw circle-cutting attachment to cut a 13-in. dia. hole. I began and ended the cut at the 1/2-in. starter hole.

I then cut a relief gap about 1/8 in. to 3/16 in. wide from the far end of the rectangle to the middle of the circle. The relief gap allows me simultaneously to clamp the template wheel at the correct angle, and to secure the cradle to the stair stringer.

Next, I attached 2x3 fences to either side of the cradle with 1-in. bugle-headed drywall screws. I used a piece of the stringer material to align the guides as I screwed them to the cradle.

It took some trial-and-error experimentation to get a correctly fitting template wheel. The tolerances are close, and the cutout left over from making the cradle was a bit too small to fit properly. I increased the radius of my jigsaw circle-cutting attachment by 1/16 in., and got the right diameter wheel. When you finally achieve a correctly fitting template wheel, make several. Extras can be used for other tread dimensions.

To cut a precise slot in the template wheel I first placed the blank wheel inside the cradle and clamped the assembly to a suitable stringer board (photo next page, left). Then I scribed a centerline through the wheel and aligned it parallel to the long sides of the cradle. I used small nails to attach fences parallel to the centerline. The space between the parallel fences must be figured exactly since the fences will guide the router as it mills the template. This stair uses treads that are called 5/4 stock by the lumberyard, but they net out at about 1 1/16 in. thick. Since I was using a 3/4-in. bit to make my mortises, I had to gain 5/16 in. to arrive at the right mortise width. That meant that the space between the parallel fences had to be the width of the router baseplate plus 5/16 in.

To dimension the tread width, I scribed the outline of the tread on the template blank (photo next page, center) and cut to it freehand (photo next page, right).

Matching the treads—Most stock stair-tread material is rounded over on one edge to form the nose. On the oak tread stock we used, the factory-rounded edge could be used as it was, but we had to mill the unfinished edge with a 3/8-in. roundover bit to each side. This created an edge that fits perfectly into a mortise cut by a 3/4-in. dia. straight bit. This matching of the roundover bit to the straight mortising bit will occur again when the stair handrails are matched to the newel posts.

Laying out the stringers—My colleague Malcolm McDaniel came up with a slick way to determine tread layout for the router jig. The unit rise and run are drawn on the stringer using a carpenter's square in the usual way, but just for two treads—enough to measure directly with dividers the distance from nose to nose (drawing, below right). Next set your combination square

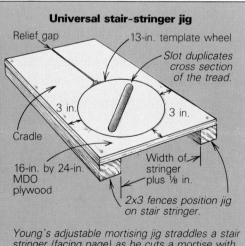

Universal stair-stringer jig

Relief gap

13-in. template wheel

Slot duplicates cross section of the tread.

3 in.

3 in.

Cradle

16-in. by 24-in. MDO plywood

Width of stringer plus 1/8 in.

2x3 fences position jig on stair stringer.

Young's adjustable mortising jig straddles a stair stringer (facing page) as he cuts a mortise with a plunge router. The heart of the jig (above) is a wheel held fast inside a housing, called a cradle. The wheel adjusts to any stair layout, and the entire assembly is secured to the stringer with a bar clamp and a pair of shim shingles.

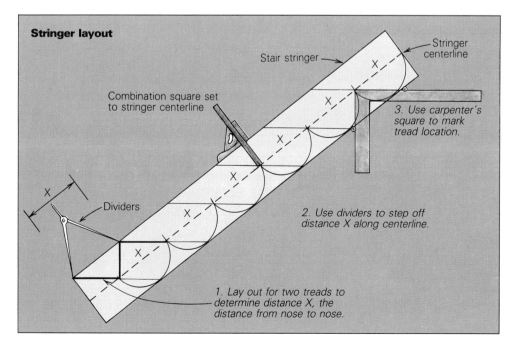

Stringer layout

Stair stringer

Stringer centerline

Combination square set to stringer centerline

3. Use carpenter's square to mark tread location.

Dividers

X

2. Use dividers to step off distance X along centerline.

1. Lay out for two treads to determine distance X, the distance from nose to nose.

Making the template. To cut the template, Young first draws a centerline through the wheel, left, and aligns it parallel with the long edges of the brake. Next he tacks a pair of parallel fences to the cradle to guide the router as it cuts the template. Young uses a section of the tread material to scribe its dimensions onto the wheel, center. Then he routs freehand to the scribe marks, right.

for half the width of the stair stringer. Holding the combination square against one edge of the stringer with one hand, swing the dividers with the other to step off the nose-to-nose increments down the middle of the stringer. At every point marked by the dividers, use the carpenter's square to mark the location of each tread.

To cut a mortise, place the jig onto the stringer and rotate the template wheel to the first run line. Now clamp the jig securely to the stringer and recheck the alignment. You may line up either side of the slot to the tread line so long as you are consistent for the entire stringer. To avoid confusion, I make a mark on the side of the line that I intend to mortise. Once I'm satisfied with the alignment, I carefully plunge my mortising bit through the slot in the template and into the workpiece.

I cut ¾-in. deep mortises for this stair. Each one needed two passes with the plunge router—a ⅜-in. depth setting followed by a ¾-in. setting. Because you will need tail pieces to clamp the jig, the stringer should not be cut to length until the last mortise is routed.

Assembling the treads and stringers— After the treads are rounded over and cut to length, they can be let into one of the stringers that has been laid on the floor, mortise side up. Our fit was snug enough to require a gentle assist with a hammer, interceded, of course, by a piece of scrap. We placed the second stringer atop the standing treads and tapped it home, starting at one end and working little by little to the other end. We didn't use any glue because there is little lateral force on the stair. Instead, we predrilled the stringers and counterbored at each mortise for three 2¼-in. bugle-headed screws—primarily to keep the stair together during installation. On the exposed stringer, the screw holes are filled with hardwood plugs.

Newel posts—We made our newel posts from clear, kiln-dried Douglas fir 4x4 stock, reduced to 3 in. by 3 in. for a slimmer profile. Each post is secured by a ⅝-in. threaded steel rod, which

is screwed into a steel flange mounted to the floor (drawing, facing page, bottom left). I began making the newel posts by sinking a deep counterbore into one end of each 4x4. I used a 2-in. Planetor bit (Rule Industries, Planetor Div., Cape Ann Industrial Park, Gloucester, Mass. 01930) for this operation because it cuts well in end grain. I chucked the bit in a floor-standing drill press and mounted the 4x4 stock on a stop-jig that kept it vertical and resisted the bit's torque. I used the same arrangement to drill a shallower, smaller-diameter counterbore in the opposite end for the floor-mounted flange nut.

The next step was to rip the 4x4s down the middle with a bandsaw and surface the sawn sides smooth on a jointer. Then I routed a groove down the center of each half with a ¾-in. core-box bit, connecting the counterbores at each end. During all of this milling, I was careful to keep track of the matching halves. Milling completed, I glued and clamped each pair back together. After removing the clamps, I brought the still oversized newel posts down to their final dimension of 3 in. by 3 in. on a planer, removing equal waste on every side so that the rod hole remained centered in the post. The result of this procedure was a solid-looking newel post, almost indistinguishable from a post that hadn't been cut in half and reassembled.

I had the mounting plates made up at a local sheet-metal shop from ³⁄₁₆-in. steel, with a ⅝-in. nut welded to the center of each one. The plates are held in place by four 3-in. #10 screws, driven into doubled joists at each landing. The finish floor is covered with tile, so I didn't bother to mortise the plates into the floor.

Handrails and balusters—We used kiln-dried Douglas fir for the handrails and balusters. The handrails are 2x4s, rounded over on all edges. (Check building codes for design requirements.) The balusters are 1x2s. To house the balusters, I plowed a ¾-in. wide groove in the underside of all the handrails, and a corresponding groove on the top edges of the stringers. The balusters are positioned and secured by 1x spacers.

Mortising the newel posts—Each newel post is mortised to receive both the handrail and stringer. I cut both these mortises with the same kind of template that I used on the stringers, but I used a 1-in. dia. bit to match the ½-in. round-over on the handrails. Before I made any cuts, I temporarily installed the newel posts and made sure that they were plumb. This let me accurately measure the length of the handrails, and determine the plumb-cut face of both the handrails and the stringers.

The mortises in the newel posts are ½ in. deep, and the handrails are glued in place. The sharp angles at the end of the handrails and stringers have to be squared off so the assembly will fit properly (drawing facing page, top right).

Assembly—After threading the steel rods into their floor-mounted nuts, we lowered each newel post over its rod. Then we glued the stringers and handrails into their mortises, and cinched down the posts with a deep socket wrench. Next, we inserted the balusters into the slots on the top of the stringer and the bottom of the rail, alternating spacers and balusters and gluing the entire assembly. We topped the newel posts with an oak cap.

After cleaning up excess glue and sanding, we varnished the entire stairway with four coats of satin marine spar varnish. The result is a clean, spare and surprisingly strong stairway (photo facing page). The universal stair-stringer jig did what we needed it to do. It guided the router bit with precision, and it reduced the time spent preparing each stringer to less than is spent cutting a conventional stringer. □

Bill Young is a contractor in the San Francisco Bay Area.

The finished stair (facing page) is a light yet strong assembly of mortised parts that lets light and air into the stairwell. Except for the newel posts, which have been planed to 3x3s, all of the stair parts are made of dimension boards straight from the lumberyard.

From *Fine Homebuilding* magazine (August 1985) 28:38-41

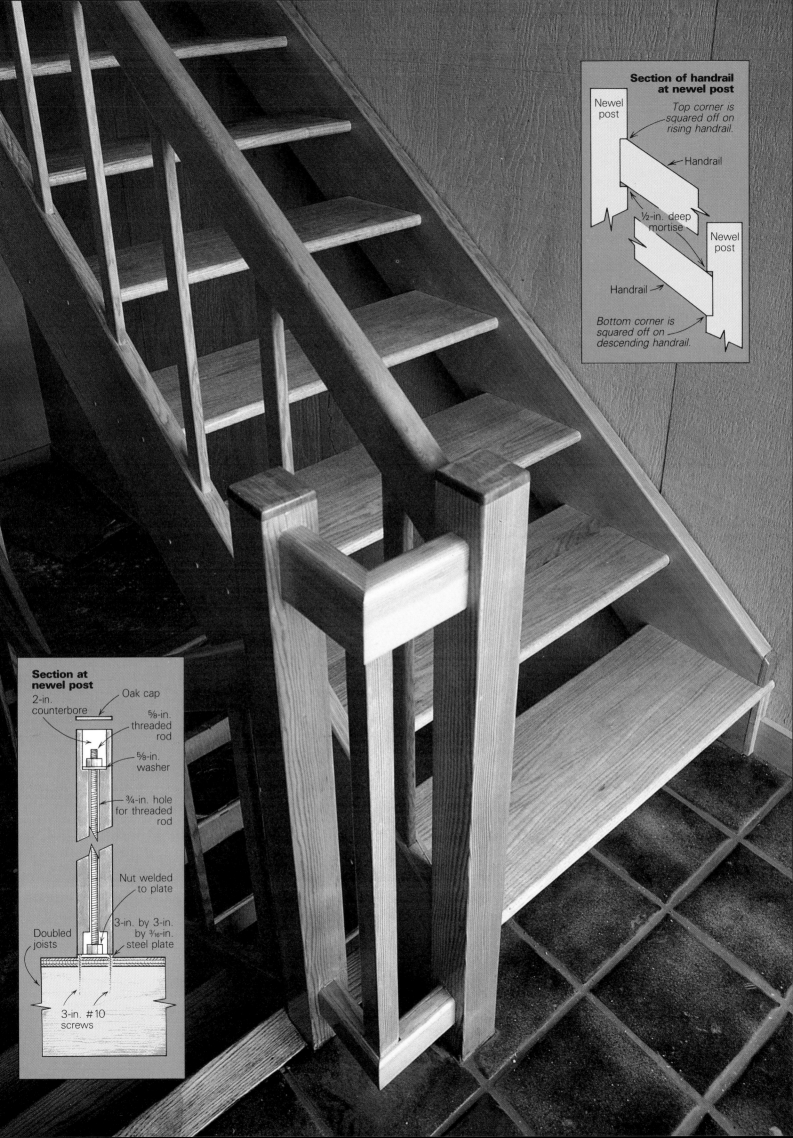

Section of handrail at newel post

Newel post

Top corner is squared off on rising handrail.

Handrail

½-in. deep mortise

Newel post

Handrail

Bottom corner is squared off on descending handrail.

Section at newel post

2-in. counterbore

Oak cap

⅝-in. threaded rod

⅝-in. washer

¾-in. hole for threaded rod

Nut welded to plate

Doubled joists

3-in. by 3-in. by 3/16-in. steel plate

3-in. #10 screws

Building a Helical Stair

Laying out a spiral stringer
with a little help from the trig tables

by Rick Barlow

A house I recently finished near Telluride, Colo., needed a spiral stairway to wrap halfway around its 4-ft. dia. stone chimney. I decided that a helical stringer supporting the treads by fabricated steel brackets would satisfy structural requirements and complement the house's contemporary design.

In theory, making a laminated wood helix isn't that difficult. Basically all you do is wrap successive layers of wood around a cylinder so that it spirals upward at a constant angle. But to glue up a helical bent lamination like the one I wanted for my stair stringer would require a cylindrical bending form larger than the chimney itself. I could have framed up such a form from curved plywood plates and 2x studding, but that would have eaten up a lot of time and materials.

I chose instead to make a bending form from five pieces of ¾-in. plywood and a strip laminated from three layers of ¼-in. plywood (drawing, below). The trick was to space the on-edge pieces of ¾-in. plywood at equal intervals along the proposed rise of the stairway, and to locate on each of these supports—or bulkheads—a 10-in. wide plane that would lie precisely ¾ in. below the surface of

my theoretical half cylinder. To these five planes I could attach three layers of ¼-in. plywood. Glued together, this rigid plywood curve would be the platen, or base, of the bending form. To it I'd clamp the mahogany plies that would make the helical stringer.

In section, I wanted the stringer to be 5 in. wide by 6 in. deep, with its corners radiused with a ¾-in. roundover bit. Plies ¼ in. thick would easily make the required bend, and I used 20 of them 6 in. wide.

The centerline of the helical stringer would lie 42½ in. out from the centerline of the chimney cylinder. This meant that a typical 36-in. wide tread would clear the stonework by about ½ in. Subtracting half of the stringer thickness (2½ in.), gave me a 40-in. radius for the concave side of the stringer. This dimension then was the radius of the outside (convex side) of the bending form.

To create the helix, the first and fifth supports of the bending form needed to support the stringer against vertical edges. The second and fourth needed 45° edges, and the stringer would run over a horizontal edge at the middle support.

I nailed the five plywood bulkheads to the

floor with 2x4 blocks and braced them diagonally to handle the forces they'd have to withstand during gluing and clamping. Next, I bent the three layers of ¼-in. plywood over these supports. The first layer was screwed and glued to the bulkheads, and the next two glued on top. I made sure that during this process the whole assembly stayed properly aligned throughout its length.

I built the form so that its centerline would be 5 in. off the floor. This way the stringer could run long at each end and be trimmed to fit the floor and rim joist later. I covered the plywood helix with a strip of visqueen so the mahogany wouldn't get glued to the form.

Laying out the stringer—The laminations for the stringer started out as 16-ft. long mahogany boards. I had them milled to a thickness of ¼ in. and ripped into 6-in. widths. To determine the length of these pieces, a little trigonometry was required.

First, I drew the stringer in elevation as if it were a straight piece of wood (drawing, facing page). The height of the triangle thus formed was the rise of the stringer. This is the total rise (distance from the first floor to the second) minus one riser height, as in a standard staircase. The total rise (110 in.) breaks down into 14 rises of 7.86 in. (about 7⅞ in.). Therefore the total rise minus one rise (7⅞ in.) for the top step (landing) results in a stringer rise of 102⅛ in.

The base of the triangle represents the distance traveled by the stringer in its run around the chimney. In this stair, the stringer winds through a 180° arc around a 42½-in. radius at the center of the lamination. Its run, therefore, is π × 42½ in., or 133½ in.

The length of the stringer itself is the hypotenuse of this right triangle. The Pythagorean theorem $(a^2 + b^2 = c^2)$ tells us it is 168 in. long, just over 14 ft.

Since the outer plies travel larger radii than those at the center, they required slightly longer pieces of wood. I ordered the 16-footers to be sure of sufficient length.

The drawing of the flattened helix also helped me determine various angles. By finding the cotangent of the triangle (cot A = adjacent/opposite = 133.5/102.125 = 1.3072), I got the pitch angle of the stringer, 37° 25′.

The complement of this angle is 52° 35′. This is the angle that the stringer makes when intersected by a plumb line, an angle I needed

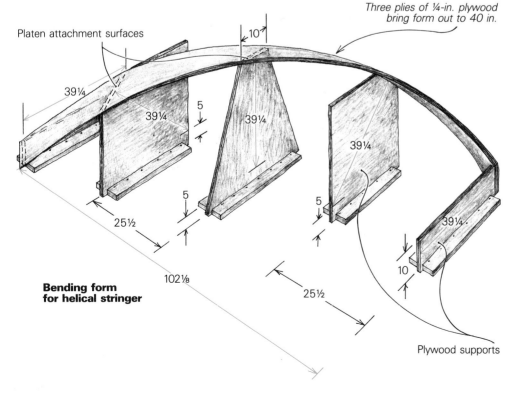

*Three plies of ¼-in. plywood
bring form out to 40 in.*

Platen attachment surfaces

10

39¼

39¼

5

39¼

39¼

5

5

25½

5

39¼

10

39¼

102⅛

25½

**Bending form
for helical stringer**

Plywood supports

Photos: Rick Barlow; Illustrations: Christopher Clapp

to know later when I counterbored the underside of the handrail to receive its posts.

I laid out the treads on the drawing of the straightened helix, and divided the 102⅛-in. rise of the stringer into 13 equal rises of about 7⅞ in. each. I figured the distance along the stringer from one tread to the next by dividing 168 (the stringer's length) by 13. This distance is 12.96 in.

Gluing up—Now, with the mathematics out of the way, I started gluing up the stringer. Since I wanted the glue to stick to the steel tread-support plates as well as to the wood, I used a resorcinol-formaldehyde glue. It's a hassle to use, and almost impossible to remove from tools and hands when it sets up, but it's strong stuff.

After spreading the glue with a small roller, I laid three or four 16-ft. long mahogany plies on the upright center support, clamped them there and then bent each side down to the ends of the plywood helix. You need lots of clamps. I have eight bar clamps, four handscrews, four smaller bar clamps and ten C-clamps. I borrowed about this many again from a friend and still had what I think is the minimum number of clamps for the length of this stringer, one clamp about every 3½ in.

Resorcinol glue has to cure at 70°F or above. It will set faster at temperatures higher than this, but it won't bond properly at temperatures below 70°F. Since I was doing this laminating in a house under construction in the wintertime at 9,000 ft. in the Colorado Rockies, I needed to camp out at the job site for seven days and nights to make sure the correct temperature was maintained while the laminating was going on. I glued up three or four layers at a time, letting them dry overnight before removing the clamps.

To rout the slots that would receive the steel tread-support plates, I laminated the stringer without glue between the third and fourth laminations and between the 17th and 18th laminations. This let me separate the first three and the last three plies from the center core of the stringer, rout out slots for the plates, insert and secure the plates, and then glue these inner and outer laminations onto the center one. The entire stringer was shaped and sanded before the plates were inserted so the steel would not get in the way.

I made two guide templates for my router to rout out the slots. One was for the thirteen slots on the convex curve of the stringer, and the other was for the thirteen slots on the concave curve of the stringer. Each jig was made to rout a 3-in. wide slot ¼ in. deep. To locate and orient each slot, I hoisted the center core lamination of the stringer into the position it would ultimately take, trimmed off the ends to fit, and temporarily secured it there

Barlow's helical stairway winds its way around a 4-ft. diameter stone chimney. The single stringer was laminated from 20 plies of ¼-in. mahogany. The treads, made of plywood and solid cherry end boards, are supported by brackets fabricated from ¼-in. steel plate.

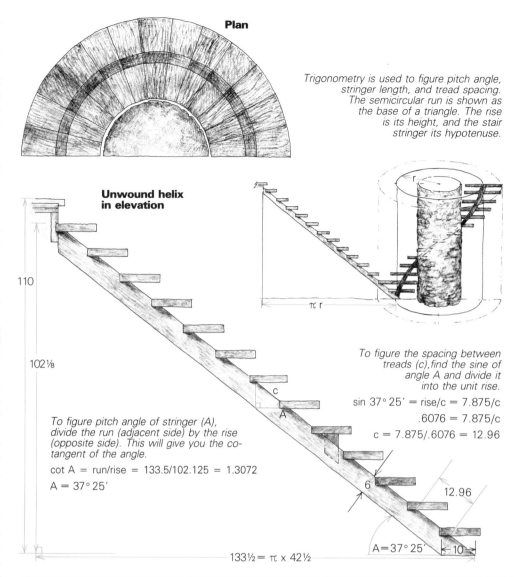

Plan

Trigonometry is used to figure pitch angle, stringer length, and tread spacing. The semicircular run is shown as the base of a triangle. The rise is its height, and the stair stringer its hypotenuse.

Unwound helix in elevation

110

102⅛

To figure the spacing between treads (c), find the sine of angle A and divide it into the unit rise.

$$\sin 37°25' = rise/c = 7.875/c$$
$$.6076 = 7.875/c$$
$$c = 7.875/.6076 = 12.96$$

To figure pitch angle of stringer (A), divide the run (adjacent side) by the rise (opposite side). This will give you the cotangent of the angle.

$$\cot A = run/rise = 133.5/102.125 = 1.3072$$
$$A = 37°25'$$

c

A

6

12.96

A=37°25'

10

133½ = π × 42½

Before its treads are attached to the steel-plate brackets, the stairway is installed. The temporary midway support will be replaced by a steel pipe that will be grouted into a masonry planter below.

laminating two ½-in. thick pieces onto a ¾-in. thick core, yielding a total thickness of 1¾ in.—the same as the cherry. The ¾-in. middle ply of plywood was cut 1 in. shorter on each end, to form a ¾-in. by 1-in. groove. I cut tongues on the cherry end boards and screwed and glued them to the center pieces.

I bandsawed the ends of the treads to conform to the semicircular plan view, and rounded over all the edges with a ½-in. piloted router bit. Finally, I fastened each tread to its mounting plate with four lag bolts.

Bending the railing—The form I built to laminate the handrail was like the one I used for the stringer, only with a larger radius. The handrail would be just under 16 ft. long, so I could use the same boards. I used eight laminations ¼ in. thick and 3 in. wide, making the handrail 2 in. by 3 in. in section. To accept the handrail posts, I counterbored two ½-in. dia. holes in each tread at the centerline radius of the handrail, and spaced them evenly 7¾ in. apart. I then counterbored the underside of the handrail. These holes had to be drilled at an angle of 52° 35′ (the complement of the pitch angle). To find the proper spacing for

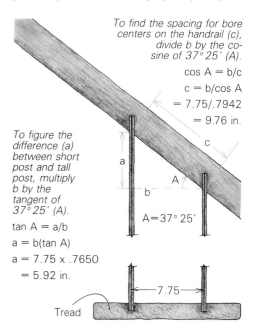

To find the spacing for bore centers on the handrail (c), divide b by the cosine of 37° 25′ (A).

$$\cos A = b/c$$
$$c = b/\cos A$$
$$= 7.75/.7942$$
$$= 9.76 \text{ in.}$$

To figure the difference (a) between short post and tall post, multiply b by the tangent of 37° 25′ (A).

$$\tan A = a/b$$
$$a = b(\tan A)$$
$$a = 7.75 \times .7650$$
$$= 5.92 \text{ in.}$$

A = 37° 25′

Tread

7.75

the bore centers here, I divided 7¾ (the horizontal spacing) by the cosine of the pitch angle, as explained in the drawing. Then ½-in. dia. steel rods, painted flat black, were inserted in the treads. Since each tread gets two rods, one must be longer than the other. By multiplying the tangent of the pitch angle by 7¾, I found the regular difference in length between the long and short rods.

With the help of friends, I worried the rods into their holes one at a time, starting at the bottom rod and working up to the top. By using a rubber mallet and some clamps, I brought the handrail down onto the 26 rods.

All that remained was to carpet the plywood center pieces and metal tread plates. Since the treads are visible from below, this was a difficult job, but it turned out fine. □

Rick Barlow is a contractor in Colorado.

with clamps and a 2x4 leg at its midpoint. Then using a level, I drew plumb lines on each side of the 26 slots. This way I was sure they would be plumb in the finished staircase.

These slots were 3 in. wide, and their sides followed the curve of the stringer, so the legs of the plates also had to be slightly curved. I hit the middle of each leg with a sledge while it rested on two other plates along their edges. One or two whacks did the trick.

I drilled holes in these plates for four 8d nails to hold them onto the stringer. These and the resorcinol-formaldehyde glue made for a solid bond between metal and wood.

With the tread-support brackets installed, I put the stringer back onto the form and glued the inner and outer laminations onto the core. They were already shaped and sanded, so I had to align the pieces precisely, and use pine blocks to keep the soft mahogany from being crushed by the clamp jaws. I was also careful to spread the glue so not much would ooze out of the joints. What squeeze-out there was, I wiped off before it dried. This assembly was left to dry overnight (photo above).

Installing the stringer—After a little more finish sanding, I cut the outer plies to fit the floor and rim joist. A 180° helix like this one

needs support at its midpoint. I drilled a 2-in. hole underneath the stringer and inserted a galvanized pipe that sits discreetly in some rockwork below.

When the stringer assembly was firmly attached in place, the horizontal tread plates were welded to the vertical plates protruding from the stringer. Using a universal level, I held the horizontal plates level in both planes, while a welder attached the plates from underneath. After cleaning up the welds with a file and wire brush and before attaching the treads, I oiled the stringer and painted the plates and the support pipe flat black.

Making the treads—Working out the dimensions of the tapered treads required more figuring. The radius to the outside of the treads was 60½ in. Multiplying by π gives the length of a 180° arc, or 190 in. Dividing this by 13 treads gives 14.62, so the outside arc of each tread is 14⅝ in. Similarly, I found the length of the inside arc of each tread to be 5¹⁵⁄₁₆ in. Nosing added an inch, so each tread is 36 in. long with a 15⅝-in. arc at its outer end and a 6¹⁵⁄₁₆-in. arc at its narrow end.

I planned to carpet the centers of the treads, and leave solid cherry end boards on each end. I made a center of AC plywood by

From *Fine Homebuilding* magazine (August 1983) 16:70-72

Sculpted Stairway

Imagination and improvisation coax an organic shape from 42 plies of Honduras mahogany

by Jody Norskog

My brother Noel and I first met Chris Bradley and Steve Zoller in Santa Fe, N. Mex., in the summer of 1980. They were researching solar homes, looking for ways to improve a house they were designing and building, and generally checking out the Santa Fe scene. They were designing a house in Laguna Beach, Calif., and mentioned that they might have a stairway for us to build. We said sure, keep us in mind, and never expected to see them again.

Six months later we got a call from Laguna Beach; "Interested in doing that stairway?" Bradley and Zoller had been working on a design, but they weren't satisfied with the way it looked and wanted us to help.

Design concepts—Bradley and Zoller did not want a stairway framed up in the conventional fashion, and were unwilling to settle for a commercial spiral stair. They wanted a stairway that would be sculptural as well as functional—something that would be enjoyable to look at and use, but would not block the view from a large glass area.

Noel and I began debating design, materials and details. Originally we considered building the entire stairway out of steel. We later decided on wood after eliminating some other possibilities. After building models to study forms and detailing, we began exchanging ideas between Santa Fe and Laguna Beach.

The design we finally agreed on was far from conventional. We chose to support the treads and handrails with a single beam, which could be bent to form a compound curve. Imagine taking a tree trunk and bending it to the desired curve. And imagine notching the trunk and fastening a plank at each notch to make treads, and attaching a handrail on each side of the treads. This was the basic concept of the project. Our trunk would be a bent lamination.

Honduras mahogany was being used for the

Jody and Noel Norskog are partners in their firm, Norskog and Norskog. They build furniture and do architectural woodwork.

Looking as if it grew there, the Norskogs' mahogany and steel stairway winds its way gracefully to the second story. A mammoth undertaking, this stair took three months' work in their Santa Fe shop, and another six weeks at the Laguna Beach site.

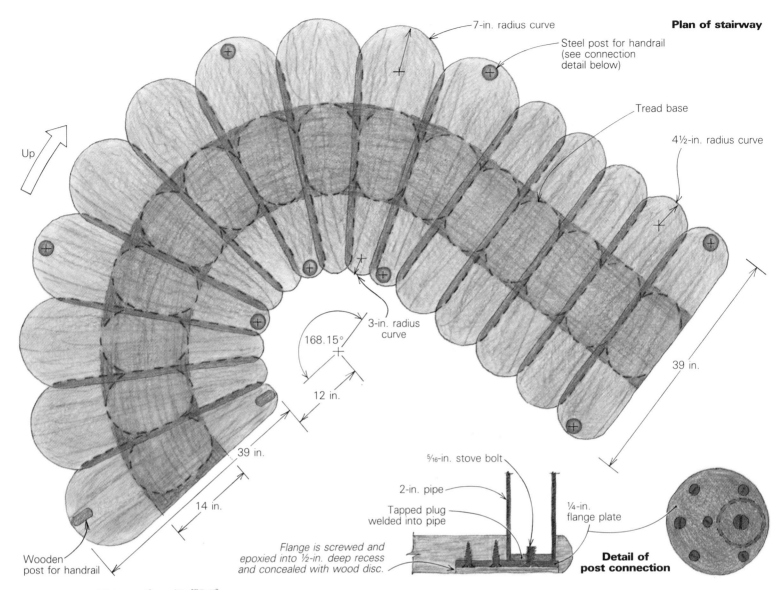

7-in. radius curve

Steel post for handrail
(see connection
detail below)

Tread base

4½-in. radius curve

Up

168.15°

3-in. radius
curve

12 in.

39 in.

39 in.

14 in.

Wooden
post for handrail

⁵⁄₁₆-in. stove bolt

2-in. pipe

Tapped plug
welded into pipe

¼-in.
flange plate

*Flange is screwed and
epoxied into ½-in. deep recess
and concealed with wood disc.*

**Detail of
post connection**

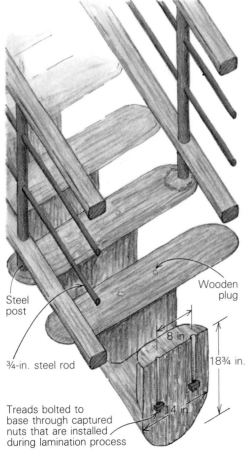

Steel
post

Wooden
plug

¾-in. steel rod

8 in.

18¾ in.

14 in.

Treads bolted to
base through captured
nuts that are installed
during lamination process

wood details and trim in the house, and it made sense to use it for the stairway as well, even though the wood had some bad qualities for the job at hand. While its crisp tissue is easy to carve and shape, it's not an easy wood to bend.

Once the design of the stairway had been reasonably defined, we worked with a structural engineer to determine what it would take to hold the thing up. After producing 30 pages of calculations, he was able to give us the information we needed—how to size the cross section of the trunk and how to design and size the hardware for connecting the trunk to the first and second floor.

The design and engineering process had stretched out over nine months before we got the go-ahead to build. Finding the wood for laminating the trunk turned out to be as hard as any other part of the project. No supplier in New Mexico could get what we needed. By doing some sample bending we determined the maximum thickness for each ply (or lamina) was ⁵⁄₁₆ in. Anything thicker was either too difficult to bend or too prone to tension failure—cracking on the convex side of the curve.

The problem was where to find pattern-grade Honduras mahogany boards ⁵⁄₁₆ in. thick by 15 in. wide by 18 ft. long. (The term *pattern grade* is peculiar to mahogany. It simply desig-

nates clear, stable stock, suitable for making foundry patterns.) Planing 4/4 stock down to ⁵⁄₁₆ in. would involve criminal waste, and re-sawing thicker, 18-ft. long boards was beyond the capability of our 14-in. bandsaw. We had to find a supplier who had lots of mahogany in the right dimensions. After much research and many telephone calls we located what we needed in a lumberyard in Long Beach, Calif. The lumberyard resawed 8/4 stock and surfaced the planks to our ⁵⁄₁₆-in. thickness. It was ironic that we bought the wood in southern California, had it shipped to Santa Fe, worked it, and then returned it to southern California.

Gluing up the trunk—The bending form for laminating the beam was an elegant piece in its own right (photo facing page). At either end there were plywood end plates with 2x4 platens for the actual working surfaces of the form, spanning between the plates. Essentially, it was like building a 2x4 stud wall to match the inside curve of the beam, and then leaning it on its side. Each 2x4 served as a clamping platen, and as a reference point for locating the position of the riser for each step. One person walked into the shop and said, "What are you guys building, an airplane?"

Few of the boards we got were wide or long enough, which meant that we had to butt sev-

Photos: Noel and Jody Norskog; Illustrations: Christopher Clapp

Work on the staircase began with milling the stock and gluing up the 42 plies for the large lamination that would be sculpted to become the stringer or trunk for the treads. The bending form, though a simple jig made from 2x4s and plywood, had to be precisely built to produce the correct curve and twist. Because plastic-resin glue takes eight hours to cure, it took 42 days to glue up the whole lamination.

eral boards together to make up each layer of laminations. We were careful to stagger the joints from one layer to the next.

After getting the clamps, wood, and form ready, we did a dry-clamping run and felt we were ready for the first lamination. We mixed up the plastic-resin glue, applied it to the boards with paint rollers, put the wood on the form and clamped two plies together.

The next day we pulled the clamps off but the glue didn't hold. We tried it again with new boards. The same thing happened. At this point we were beginning to get frantic. Bad glue? Wrong kind of glue? Should we remill the wood? Then we discovered that plastic-resin glue requires a temperature above 70°F to cure. This was January at 7,000 ft. in northern New Mexico, where it gets cold. Our passive-solar shop maintains a comfortable temperature fairly easily—but not always 70°F. With our new kerosene heater we made several more tries, all the time refining our gluing procedure. At last we were able to get a bond that we felt good about.

Each tread was to be held in place by two countersunk bolts threaded into correspond-

ing nuts embedded into the beam. The nuts were captured in the beam as it was being laminated. After the thirteenth layer, the first set of nuts was inserted. The 2x4 platens allowed us to position each nut accurately. We routed slots for the nuts and corresponding slots for bolt access. After each nut was set in the beam, the slots were taped over so they wouldn't fill with glue, and the laminating process continued. At the thirty-first layer, the second set of nuts was embedded into the beam. We used rectangular nuts made from ½-in. bar stock instead of standard hex nuts, to prevent them from turning.

Clamping time for Borden's plastic-resin glue is eight hours at 70°F. This meant that the whole lamination (42 plies in all) would grow at the rate of one ply per day. The pipe we were using for clamps could exert only enough pressure to hold down two layers of the mahogany over the curve. (We tried once to clamp three plies, but the force required was so great that the threads on the pipes began to strip.) The first ply got the glue; the second we used as a caul to even out the clamping pressure. Each caul ply became the

following day's gluing layer. Occasionally the cauls would fail in tension. There seemed to be no way of determining which board would fail except by testing it. Each day meant pulling the clamps off, sanding the entire surface, fitting boards for the next lamina, mixing and spreading glue, and clamping on the next ply. One lamination a day for 42 days.

Carving the trunk—Once the forty-second lamination was clamped down, it was time for a celebration. Even after it was notched and carved, the glued-up trunk weighed 2,000 lb., so it weighed considerably more at this point. We rented a 3,000-lb. capacity hoist for moving the beam around. Used for pulling out automobile engines, it picked up the beam, bending form and all. Once the entire thing was in the air, it was easy to disassemble the form, piece by piece.

The location of the treads was marked by the 2x4s on the form. But after all the days of clamping, the jig was pretty well abused, and we couldn't trust its accuracy. With the assistance of the hoist, we placed the beam in upright position. By drawing the plan view on

From *Fine Homebuilding* magazine (August 1983) 16:73-77

Once the lamination was complete, notches for the treads were roughed out with an electric chainsaw (top left). Then the tread-bearing surfaces were cut flat and level with a router attached to a plywood base (bottom left). With guide strips nailed to the trunk at the layout lines, it was easy to slide the router back and forth to mill the surfaces flat. To secure the stair to the first floor, a steel plate was bolted through to the framing. The bottom of the trunk was notched and relieved to receive a steel sleeve (above), which fits into the plate and is welded to it. The top of the stair is secured in similar fashion to the second floor.

the shop floor, we were able to transfer vertical riser locations up to the beam with plumb bobs and mark the horizontal cutting lines with levels.

With the rise and run determined, we used an electric chainsaw to rough out the notches for the treads to within a ½-in. of our layout lines (photo above left). To mill the horizontal surfaces of the notches flat, Noel made a long base for the router, attached guide boards to the trunk and surfaced the notches flat with a ⅞-in. straight bit, as shown in the photo below left. To give the massive trunk sculptural relief, we carved it to look like a tube, with the risers as intersecting tubes. The bulk of the carving was roughed out with the chainsaw, then cleaned up with various chisels and rasps. This exposed a pleasing lamination pattern, and led a number of people to ask how the risers were attached to the beam.

Making the treads was the only easy part of the project. Cut from 8/4 stock, they were laid out and arranged according to grain and color, and then planed down to a thickness of 1½ in. Their ends were rounded on the bandsaw and their edges routed to a half-round. The treads were then bolted to the trunk using the captured nuts that we had glued into the beam while it was being laminated.

To hold the stairway in place, our engineer called for a 10-in. square steel sleeve to be inserted into each end of the trunk. This was no problem at the top of the trunk because it was straight. But the curved bottom required a curved rectangular steel sleeve. We fabricated one (photo above) by heating four ¼-in. steel plates and bending them to the correct curvature before they were welded.

To avoid having to make separate bending forms for the handrails, and to get more accurate results, the Norskogs clamped L-shaped brackets to each tread at the location of the handrail (left), and then clamped the laminae to the brackets. Tubular-steel rails are bent to shape and then cut to length (top right). Then each rail is fitted between the posts and welded in place (above right) between the top and bottom rails of wood.

Handrail—The balustrade was designed to complement other handrails in the house—a wooden rail at top and bottom, with two intermediate ¾-in. steel-rod rails. The balusters we made from 2-in. steel pipe. Like the trunk, the wooden handrails are bent laminations. But instead of constructing a special bending form for gluing them up, we used the stairway itself as a form by making L-shaped brackets, and clamping one to each tread. Then we clamped the handrail plies to these brackets to get the right compound curvature for the rail (photo above left).

For laminating each handrail, we picked out an 8/4 board and ripped it into strips thin enough to bend to the required curve and twist. We kept the strips together in the same order in which they were cut, rolled glue on each strip, and clamped them back together. This procedure ensured that the original grain pattern would remain intact, and that the gluelines would not be highly visible.

The stairway changes from spiral to straight ten steps up. This transition made for some of the most difficult parts of the project. It causes problems for the handrail not only because there is a transition from the curve to the straight, but also because there is a change in the angle of incline.

Looking at the stairway from the side view, the inside handrail has a convex transition and the outside a concave one. Instead of laminating these tricky compound shapes, we glued up stock large enough to carve solid sections as transitions between the curved and straight handrails.

Delivery and installation—This was also part of the job (a bigger part than we bargained for), so we rented a truck with an enclosed box. Loading the 2,000-lb. trunk was no problem with four men and a hoist on casters. We were able to roll the trunk out the shop door right into the waiting truck.

Unloading was a different story. The house where the stairway was to be installed is on one of the world's steepest habitable hillsides. And there's no easy access, just a long walk almost straight up. We lost sleep for nights trying to figure out a suitable way to get the trunk off the truck, up the hill and into the house. When we arrived at the building site,

we mustered all the men who were working on the house, and asked them to help haul the mahogany behemoth up the hill. We didn't tell them it weighed a full two thousand pounds. There was a lot of moaning and groaning, but we got it in place at last with nothing more sophisticated than grunt consciousness and brute force.

The installation went smoothly, but it took longer than we expected, as installations sometimes do. With the help of a welder, we set our mounting plates on the first and second floors (photo facing page, top right). The beam was lifted into place with another engine hoist. After positioning the trunk, the steel sleeves were attached to the ends of the beam. With the trunk in place, the steel sleeves were welded to the mounting plates. Final sanding on the trunk started, and treads were bolted down. Once the treads were down it was time to fit, mount, and shape the handrail (photos above right).

We returned to Santa Fe after six weeks in Laguna Beach. The stair was complete except for final sanding and finishing. The mahogany got six coats of Watco oil finish. □

Concrete Spiral Staircase

A massive stair made by casting treads in a precision mold and bending thin-wall tubing

by Dennis Allen

The stair's concrete treads are threaded on a steel column and locked in place by steel balusters. Each of the 300-lb. steps was cast individually in a mold lined with plastic laminate.

This circular staircase was inspired by the stone stairways built in Europe during the Middle Ages. Designer Paul Tuttle wanted to create the sense of timeless solidity that massive stone steps evoke, and the two-story greenhouse in the Douglas residence overlooking Santa Barbara, Calif., gave him his chance. The room needed both a stairway and a sculptural focus, so Tuttle captured the medieval aura with the 10-ft. high, 7-ft. dia. concrete spiral stair shown in the photo at left.

Even before I had seen the design, Tuttle asked me if I would be interested in building this staircase. My first impulse was to say no. A poured-concrete spiral stairway seemed impossibly difficult. But once I saw his drawings, I got excited by the challenges its construction presented. Along with two of my associates, I agreed to build it. But I still wasn't really sure I could.

Layout—The first thing we had to do was to determine the number of steps, the rise of each step and where the beginning and end of the spiral would fall. We eventually decided on a landing and 15 steps of 7½-in. rise. Each one is a 25° segment of a circle.

Next we made a full-scale mechanical drawing of one of the steps, including sleeves for holding the balusters, a cutout for toe space, indentations for carpeting and a sleeve for the center column. This drawing (facing page, left) proved to be an indispensable reference in all stages of the project.

Making the concrete form—After we figured out what one of our concrete treads had to look like, we had to build a form that would duplicate it 15 times. The form had to be durable, yet flexible enough to be taken apart after each pour and reassembled for the next one. The project's carpenter, John Bunyan, and I thought about our options and eventually came up with the form shown in the drawing on the facing page, right.

We used wood and plastic laminate to make the form, the main components of which were the base, the sides, the pivot end and the outer end. The base is a 3-ft. by 4-ft. piece of ½-in. plywood that we covered with plastic laminate. We made the

From *Fine Homebuilding* magazine (April 1984) 20:50-53

Steelwork

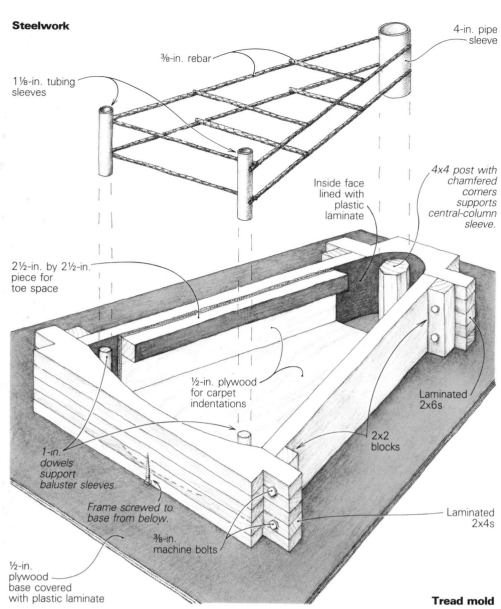

3⁄8-in. rebar

4-in. pipe sleeve

1 1⁄8-in. tubing sleeves

Inside face lined with plastic laminate

4x4 post with chamfered corners supports central-column sleeve.

2½-in. by 2½-in. piece for toe space

½-in. plywood for carpet indentations

Laminated 2x6s

2x2 blocks

1-in. dowels support baluster sleeves.

Frame screwed to base from below.

3⁄8-in. machine bolts

½-in. plywood base covered with plastic laminate

Laminated 2x4s

Tread mold

...ase, and screwed ...wo together with 26 1½-in. screws.

Next we screwed plywood pieces to the bottom and lead edge of the form to create an indentation for carpeting. The shape of these pieces had to be carefully worked out so that the carpeting would flow from one step to the next without any offset. We planned for the bottom of the form to be the mold for the top of each step, so we could place the largest of the ½-in. let-in pieces for carpeting on the bottom of the form. Pouring the steps upside down also let us trowel the bottom of each tread—the largest and most visible expanse of concrete on each step. Finally, a 2½-in. square piece of wood 28 in. long was covered with laminate and screwed to the top edge of the form. This would form the indented toe space on the bottom lead edge of each step.

The last parts of the form were the registration pins for the steelwork—two 1⅛-in. tubing sleeves to support the rebar, and the 4-in. pipe sleeve that would fit over the central column of the stairway. We screwed two 7½-in. lengths of 1-in. wood dowel to the base near the corners at the wide end of the form (drawing, above right). At the other end, we attached a 7½-in. piece of 4x4 with chamfered corners. These pegs, carpet inserts and toe board were screwed in from the outside of the form so they could be easily released when we stripped the form from the cured concrete.

Structural steel work and balusters—We decided to cast a short length of 3-in. pipe into the slab-floor footing to act as an anchor for the 3½-in. central-column pipe. Each step would have cast within it a sleeve of 4-in. pipe. These sleeves would slip over the center column and become an integral part of the rebar assembly in each concrete tread. The rebar grid would also include the two 1⅛-in. tubing sleeves into which the railing balusters would slide and be secured, as shown in the drawing above..

The sleeves in each step are attached to one another by a matrix of 3⁄8-in. rebar. Fifteen grid assemblies were required, one for each stair, and each one had to match all the others exactly in order for the balusters and central column to fit properly.

To achieve this kind of accuracy, our steel expert, Dean Upton, welded each assembly on a jig. In a 3⁄8-in. steel plate, Upton drilled holes at the sleeve centers and tapped them for ½-in. bolts. Then he tack-welded posts to the plate. These posts had been turned to fit the inside diameter of the sleeves, and were bored to accept the ½-in. bolts. The sleeves for each tread were cut to length and squared. Then Upton bolted them to the jig and welded the rebar in place.

The posts that support the handrail are 1-in. OD steel tubing with a .083-in. wall. The sleeves into which they fit are 1⅛-in. OD tubing with a .049 in. wall, allowing a clearance of .027 in. That seemed a bit loose to us, but proved to be a necessary margin during the final stair assembly.

Upton silver-soldered a ¼-in. ring (from the sleeve material) to each baluster to make sure they ended up at the right height. Their tops were cut at the angle of the stairway, and as they were installed, the balusters were rotated to match the direction of the handrail. Once the treads were in place, we plugged the underside of the baluster holes with Bondo (a brand of auto-body putty).

Picking out the pipes—Material selection required some careful sleuthing. Our basic plan called for several pipe sizes, each to fit snugly over the next. The central column is 3½-in. pipe, schedule 40. The sleeves for the steps are 4-in. pipe, also schedule 40. Pipe goes by the inside diameter (ID) while tubing goes by the outside diameter (OD). With pipe, oddly enough, the OD is constant while the ID changes with the schedule or wall thickness. A 3½-in. pipe (schedule 40) is actually 4-in. OD by 3.548-in. ID, with a wall thickness of .226 in. So depending on the schedule, we had a choice of clearances between sleeve and column. With 3½-in. pipe and a 4-in. sleeve, there is a theoretical clearance of .026 in. But

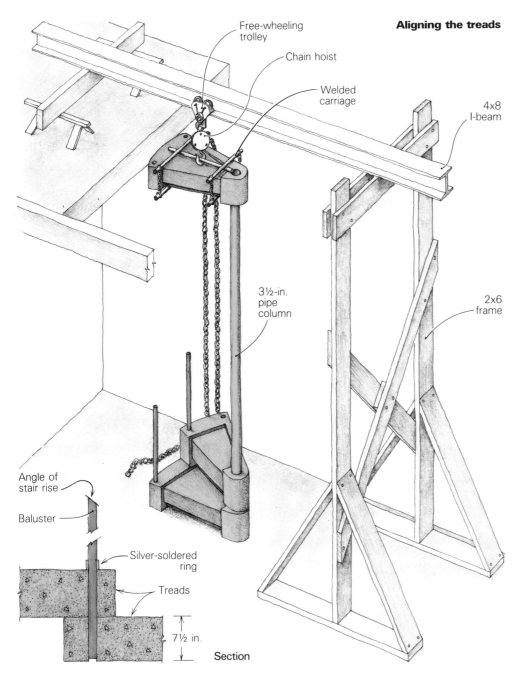

Free-wheeling trolley

Chain hoist

Welded carriage

4x8 I-beam

3½-in. pipe column

2x6 frame

Angle of stair rise

Baluster

Silver-soldered ring

Treads

7½ in.

Section

Assembly—Once the 15 treads were cast and carted to the site, our next hurdle presented itself—slipping the 300-lb. steps onto the steel column. Obviously we needed some type of device to lift the treads. It would have to be sturdy enough to carry the heavy loads, yet also adjustable so we could fine-tune the position of each tread over the center shaft.

Our solution was the homemade chain hoist shown in the drawing at left. It consisted of a 12-ft. 4x8 I-beam and a chain hoist mounted to a freewheeling trolley. We centered the beam over the column, and held it up by the stair landing on one side and a sturdy framework of 2x6s on the open end.

We fabricated a special metal carriage and harness to carry the treads as close as possible to their center of gravity. Each tread was then lifted above the 10-ft. high column, and its sleeve was centered over the shaft. Then the tread was slowly lowered into position. Once a step was in place, we would brace its outer end with a 2x4 and then one of us would tap a baluster through the aligned sleeves to secure the new tread to the one below it.

Disaster nearly befell us midway in the assembly. As we rolled the tenth tread along the I-beam, a lurch in our movements caused a sudden shift in the tread's center of gravity. Instantly the harness slipped off the carriage and the step plummeted, bouncing against several of the steps already in position and demolishing the bottom picket on its way to the floor. We were stunned. Fortunately nobody had been standing in the way of the tread when it fell. We surveyed the damage and it appeared enormous. Chunks of concrete were knocked off the treads in half a dozen places. We were so badly shaken that we packed up and went home for the day, believing the project ruined.

The next day we reassessed the damage and concluded that it wasn't as severe as it had seemed. We decided to patch the damaged edges and corners with Bondo. In some places we had to build up numerous layers of the stuff, but it worked far better than we had dared to hope. Because the Bondo was a different color from the pristine white concrete, we knew we'd have to paint the final product.

At the landing—The top step had a different shape from the others because it needed to flow into the cantilevered landing. To link the stairway to the landing, we built a triangular rebar grid to lock the top of the central steel column rigidly to the 4x12 landing girders. We welded sleeves for the balusters and the central column to this grid. The grid in turn was welded to a 4-in. by 30-in. by ⅜-in. steel plate, which was bolted to the landing framing. Then we erected a form around the grid with supports down to the floor. We were able to use several curved components from our breakdown form, but most of the pieces were new and had to be covered with plastic laminate.

We poured this last step with the same two mixes and care that we used with all the other treads, and when we took off the forms it flowed perfectly into the landing.

this clearance did not prevail on all of the pieces so some sleeves had to be turned down on the lathe.

Concrete technology—The concrete mix was critical for two reasons: weight and finish texture. In order to decrease the weight of each tread from more than 400 lb. using concrete with standard aggregate to about 300 lb., we used ½-in. Rocklite (The Lightweight Processing Co., 715 N. Central Ave., Suite 321, Glendale, Calif. 91203), a lightweight aggregate. But we also wanted a dense, pure white finish on each tread. This led to our using two different batches for each one. The outer inch or so is made up of 1 part white portland cement, 2½ parts 60-grit silicon sand and 2½ parts Cal-White marble sand (used mainly for swimming pools), made by Partin Limestone Products Inc. (PO Box 637, Lucerne Valley, Calif. 92356). Once we got this outer layer of white concrete in place, we filled the core of each tread with the lightweight mix.

We carefully measured all the ingredients

because slump was important—too much slump would cause the two mixes to flow together in the form. The lightweight mix for the core was 1 part cement, 2½ parts sand and 2 parts aggregate. To speed setting time, we added a little calcium chloride to each batch.

Originally we'd hoped to pour two steps per day, but found that producing one a day was quite an accomplishment. Placing the mixes in the form required two of us—one to tamp the outer mix and the other to keep the core mix from migrating to the edge of the form. We placed the concrete in layers, agitating it thoroughly after each layer to eliminate voids. Between each pour we cleaned the form, coated all dowels and wood insets with floor wax and sprayed the plastic laminate with silicone.

Surprise and delight filled us when we stripped the form from the first tread. The result was magnificent, but not what we'd expected. The plastic laminate made the surface smooth as glass, and a swarm of tiny, irregular air pockets made it look something like travertine. We were elated with this first success.

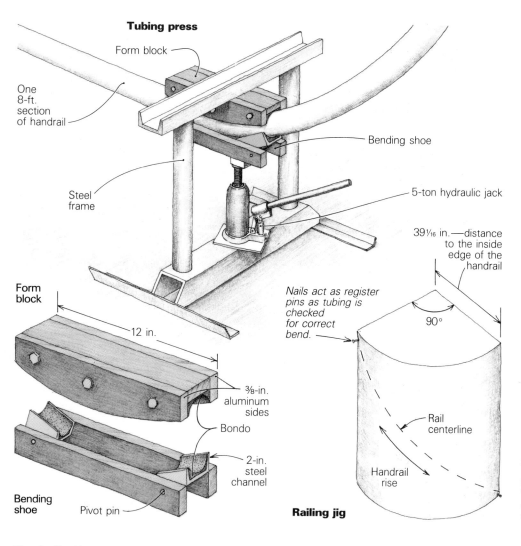

Tubing press

Form block

One 8-ft. section of handrail

Steel frame

Bending shoe

5-ton hydraulic jack

Form block

12 in.

⅜-in. aluminum sides

Bondo

2-in. steel channel

Bending shoe

Pivot pin

Nails act as register pins as tubing is checked for correct bend.

39¹⁄₁₆ in.—distance to the inside edge of the handrail

90°

Rail centerline

Handrail rise

Railing jig

Dean Upton torch-welds the handrail to a steel baluster. Although it appears continuous, the railing is composed of short segments of steel tubing that were bent on a homemade press, and then assembled on site to create the necessary helical shape.

Handrail—Probably the most challenging part of this project was bending a 1⅞-in. dia. thin-wall tube (.063 in.) into a helical handrail. We chose this size tubing because there are stock fittings for 1½-in. pipe that fit closely enough to be used with the tubing (1.875-in. OD vs. 1.900-in. OD). At the top where the stairs meet the landing, we needed a tight return bend to blend the rising stair rail into the horizontal landing rail. We made this transition with two wide-radius elbows and a little cutting and fitting. We used another stock fitting—the half-sphere cap—to finish the bottom end of the handrail, and we used floor flanges to attach the landing rails to the wall.

The radius of the stair circle was 42 in., but the radius of the line of balusters was 40 in. The inside radius of the handrail was 40 in. less ¹⁵⁄₁₆ in. (half the diameter of the tubing) or 39¹⁄₁₆ in. Taking his cue from a tubing bender, Upton designed a press that used a hydraulic jack to generate the bending force needed to arc the straight lengths of tubing. He used an oak form block with a radius of 38⅞ in., a little tighter than the required radius to allow for some springback. As it turned out, the springback was almost nil.

The forming tool we tried out first had two spools about 12 in. apart. It bent the tubing, but it also left slight dimples at each point of contact between spool and tube. A handrail with a dimple every 6 in. was totally unacceptable (it looked like a segmented worm), so Upton made a pair of bending shoes out of

channel steel and Bondo as an alternative. They needed periodic greasing to allow the tubing to slip through as it was bent. This jig, shown in the drawing above left, worked fine. A 5-ton jack supplied the pressure.

Upton first tried to form the helix as the tube was being bent by rotating the tube a little at each bend. But it was difficult to keep track of the rotation. We could calculate how much rotation was required, but to control it was tough in a small shop. Even though it wasn't the right shape for our railing, the sculpture resulting from the first try could be mounted on a stone block and placed in front of a library.

We learned two things from this attempt. One, the press could put a wrinkle-free radius in our tubing, and two, trying to form both the radius and the helix into the full-length railing was too ambitious. Instead, Upton cut the 24-ft. tube into three 8-ft. pieces. Then he made a plywood jig that had a radius of 39¹⁄₁₆ in. (drawing, above right). This jig represented about ⅓ of a turn of the staircase, and was tall enough to allow the rise of the handrail to be marked diagonally on it. As he shaped each 8-ft. section, Upton checked its bend against the jig. This worked well, and the three pieces closely approximated the required helix plus the radius.

To make final adjustments in the helical twist, each 8-ft. section was cut into three equal pieces. After tack-welding the first section to the bottom balusters, Upton rotated

the second section slightly to create the helix. Section two was then tack-welded in place, and the third piece rotated slightly more than the second and so on until all nine parts were tack-welded in place. Each piece was aligned with its neighbor by using a short offcut of 1¾ in. tube as a dowel. Before he welded the balusters to the railing (photo above), Upton torch-welded the whole unit into one continuous piece. Then all the welds were ground down, and any little pits were filled with Bondo and sanded smooth. The finished rail is painted brick red, and appears to flow as one piece from top to bottom.

Our final job was whitewashing the treads. We wanted to preserve the texture of the concrete and to have it not look painted, so we experimented with several finishes. We finally settled on white latex paint mixed with a small amount of white portland cement. This gave the surface a little roughness to the eye, but did not destroy the glass-smooth texture to the touch. One coat completely covered the grey-green Bondo, and we were done.

The project took six weeks of concentrated effort, and it kept our attention with a series of snags and surprises. But everybody is happy with the way it turned out. The stairway cost almost $10,000—a lot for one flight of stairs, but not for a sculpture that anchors a special room. □

Dennis Allen is a general contractor living in Santa Barbara, Calif.

Building a cantilevered-tread spiral stair

The retreat built on tiny Hornby Island for Jim and Judy Saks has a feeling of openness throughout. To echo this feeling, the spiral stair to the bedroom and lookout tower was designed with delicate lines, unencumbered by external support systems and limited to the tones and textures of wood. For me, each aspect of this 6-ft. dia. spiral construction was a challenge, always an adventure, and often a headache. More than once I considered selling my tools and opening a restaurant.

Layout—The central support for the staircase is a 20-ft. long yellow cedar driftwood log, about 1 ft. in dia. at the base. As a structural member of the house, the log was standing when I came on the scene, and it had to be laid out and worked in place. It was neither completely round, nor straight. Consequently, the tread-mortise positions had to be projected inward from a 6-ft. dia. imaginary cylinder symmetrically enclosing the assumed centerline of the log. Since the position of the outside end of each tread was fairly critical, individual tread lengths had to vary by an inch or two depending on the warp of the log. The entire layout for the tread positions had to be completely independent of the log, with the projected mortise positions falling arbitrarily on the log's uneven surface.

The rise of each step is 7.3 in. To climb the distance from the ground-floor landing to the bedroom landing, 13 treads travel an arc of 292° (drawing A)—one tread per 22½°. This let me find the centerline of each tread, and ignore the overlap of the treads, which was an inch on each side.

To get the mortise positions for the treads, I made a flat plywood pattern of the 292½° arc, as shown in drawing B. I drew lines on the pattern dividing it into 13 segments, and made a mark for the centerline of each segment. Then I cut out the center of the pattern so it would fit around the base of the log column, and positioned it on the floor where the first tread would start, minus the overlap. I put some index marks on the pattern and the column for future reference points, then I transferred the centerlines of each tread onto the log using a pencil and a plumb bob. I marked the rise intervals on a story stick and transferred them as top-of-tread lines to the corresponding tread centers already scribed on the pole.

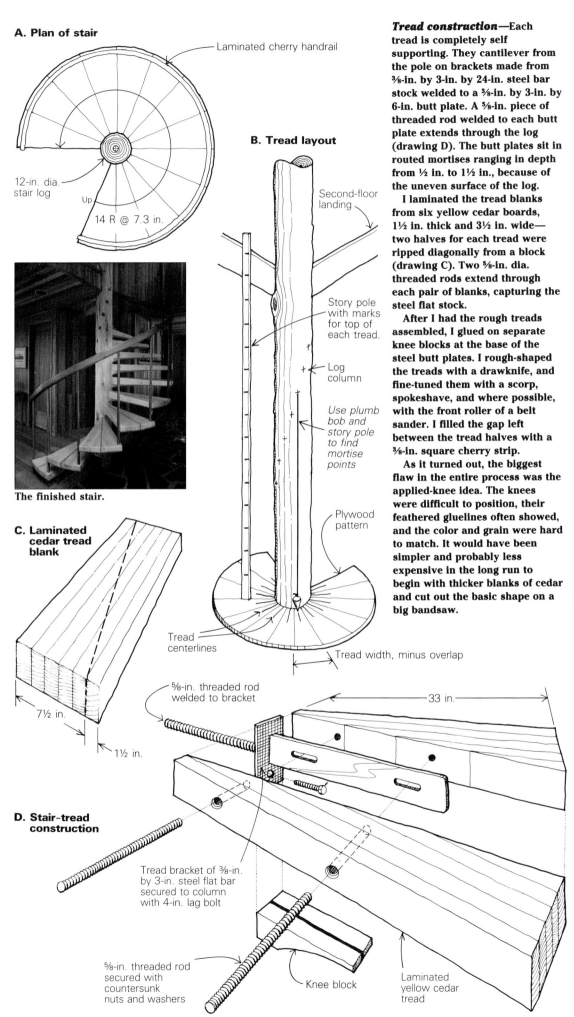

A. Plan of stair

Laminated cherry handrail

12-in. dia. stair log

Up

14 R @ 7.3 in.

The finished stair.

B. Tread layout

Second-floor landing

Story pole with marks for top of each tread.

Log column

Use plumb bob and story pole to find mortise points

Plywood pattern

Tread centerlines

Tread width, minus overlap

C. Laminated cedar tread blank

7½ in.

1½ in.

D. Stair-tread construction

⅝-in. threaded rod welded to bracket

33 in.

Tread bracket of ⅜-in. by 3-in. steel flat bar secured to column with 4-in. lag bolt

⅝-in. threaded rod secured with countersunk nuts and washers

Knee block

Laminated yellow cedar tread

Tread construction—Each tread is completely self supporting. They cantilever from the pole on brackets made from ⅜-in. by 3-in. by 24-in. steel bar stock welded to a ⅜-in. by 3-in. by 6-in. butt plate. A ⅝-in. piece of threaded rod welded to each butt plate extends through the log (drawing D). The butt plates sit in routed mortises ranging in depth from ½ in. to 1½ in., because of the uneven surface of the log.

I laminated the tread blanks from six yellow cedar boards, 1½ in. thick and 3½ in. wide—two halves for each tread were ripped diagonally from a block (drawing C). Two ⅝-in. dia. threaded rods extend through each pair of blanks, capturing the steel flat stock.

After I had the rough treads assembled, I glued on separate knee blocks at the base of the steel butt plates. I rough-shaped the treads with a drawknife, and fine-tuned them with a scorp, spokeshave, and where possible, with the front roller of a belt sander. I filled the gap left between the tread halves with a ⅜-in. square cherry strip.

As it turned out, the biggest flaw in the entire process was the applied-knee idea. The knees were difficult to position, their feathered gluelines often showed, and the color and grain were hard to match. It would have been simpler and probably less expensive in the long run to begin with thicker blanks of cedar and cut out the basic shape on a big bandsaw.

Mortising the column—For each tread I had to cut mortises for the tread butts, drill the bolt holes through the column, and counterbore for nuts and oversized washers.

I made multiple passes with a ½-in. straight-flute bit in a heavy-duty router to cut the mortises. To guide the router, I used a jig that could be positioned and secured to the column with clamps (top photo). Accuracy here was crucial because the flat bottom of the mortise formed the seat for the butt block of the tread bracket. The slightest inaccuracy in any plane would translate out to the end of the tread, putting it hopelessly out of position.

I made the jig from a piece of ½-in. plywood, 12 in. by 18 in., with 1x2 fences for the four sides of the mortise. For clamp jaws, I used 18-in. long pieces of 2x2 at the top and bottom of the jig, and opposite the column. They were linked by threaded rods. To bring the jig into plumb, I checked the bed on which the router moved for plumb in both planes, and adjusted it with wooden wedges.

Tangential orientation of the jig was the most crucial, and the most difficult. The jig's flat surface had to be perpendicular to the final centerline of the tread, and this angle had to be set from a point as far from the pole as possible in order to minimize error. To do this I added a 14-in. plywood tongue extending horizontally from the lower edge of the jig, with the tread centerline marked on it. I adjusted the jig with wedges between its base and the column until the centerline on the tongue was plumb with the corresponding center marked on the plywood pattern on the floor.

Mortises routed, I drilled bolt holes using a guide block that fit into the mortises. It both indexed the hole position and trued the ⅞-in. auger bit so that I could drill a hole perpendicular with the mortise bottom. I did my counterbores with an expansion bit. To center the spur on the bit, I tapped square wood plugs into the last 2 in. of the bolt holes, and then drilled them out during the counterboring. Leftover plug ends came out easily using a length of dowel as a drift pin.

The test—At this point the finished treads were bolted in, and the staircase was ready for a test climb. Since the railing had not yet been built, the treads were supported only by the center pole. Up to this point the

To cut the tread mortises in the column, Grunewald used this jig to guide the router. For this cut near the landing, he removed the top pair of 2x2 clamping jaws, and nailed the jig to the log column.

By clamping gusseted brackets to the stair treads, Grunewald was able to use the stair as the form for the laminated handrail. The edge of each bracket was positioned to correspond with the line of the handrail, and their upright legs served as clamping surfaces during glue-up.

engineering had been theoretical. For a practical test, I asked an unbiased (and well-insured) friend to try out the stairs and to report on three things: bounce, wiggle and ease of ascent. After a couple of trips he said that the stairs were easy enough to climb and didn't deflect much vertically, but that they wiggled unnervingly.

It was clear that the treads had to be tied together to stiffen the whole structure. The obvious solution was to use the balusters, fastening them into one tread and against the face of the next tread up. But this couldn't be done until the handrail was built.

Handrail—I laminated the rail out of nine layers of ³⁄₁₆-in. by 3-in. cherry strips, with the joints staggered at 2-ft. intervals. The staircase became the form, with gusseted angle brackets on the treads providing clamping points and forming the helical curve (bottom photo).

I had originally thought to glue up all nine layers at once. But since the weather was hot and I was workng alone I elected to do two separate gluings, one of four layers, the next of five, hoping to avoid some of the panic that glue-laminating can cause. That turned out to be a wise decision.

After it cured, we plucked the handrail from its form and planed it on the Hitachi jointer/planer. To do this, I laid the machine on its side on the porch, and fed the rail through like a huge corkscrew to a helper stationed downhill. Once it was planed, I put it back in place and marked the baluster positions. I drilled these using an angled guide block that clamped to the outside face of the rail. I cut the balusters from 1-in. square cherry blanks. Rather than turning them on the lathe, I found it quicker and easier to do four passes over a table-mounted router with a ½-in. roundover bit. I split the bottom of each baluster for a blind wedge, and filed a flat spot to seat firmly with the edge of the adjacent tread.

Each baluster was then wedged and pinned into the top face of a tread and screwed to the front edge of the tread above, effectively tying the staircase together and eliminating almost all of the movement. The predrilled rail was then tapped onto the upper ends of the balusters, and except for the many hours of sanding with 150-grit paper and applying tung oil, the job was finished. □

John Grunewald lives on Hornby Island, B. C.

Drawings: Vince Babak; Photos this page: Boh Helliwell

Outside Circular Stairway

A handsome addition without fancy joinery yields covered access to two levels

by Tom Law

In the Annapolis, Md., area, tidal creeks and rivers from the Chesapeake Bay produce some steep building lots. My clients owned a ranch-style house on just such a site. To enter the house you had to park in a space along the roadside and walk down steps onto a deck; then to reach the front door you had to walk to the far end of this deck. The owners didn't like this arrangement. The long walk to the front door was inconvenient, and the interior stairway to the basement took up valuable living-room space. An architect neighbor conceived the solution to this problem—an exterior circular stair, located closer to the road, that would provide sheltered access to both the living room and the basement. The finished stairway is shown in the photo below. I got the plans in the form of a freehand sketch. The project involved removing the interior stairs, filling in the floor, building the spiral stairway itself, and cutting a hole in the exposed concrete block wall below the deck to make a new doorway to the basement.

I started by laying out the new entry door about midway down the deck, which still allowed it to open into the living room. The right side of the door was to be the center of the new spiral-stair enclosure. Squaring over to the outside edge of the deck, I dropped a plumb bob to the ground. Using a post-hole digger, I went down about 5 ft. with a 20-in. diameter hole until I hit sandstone. I formed up an octagonal pier of the same dimension 6 in. above grade and poured concrete.

The plan called for a round central post, but I was planning simply to glue and nail the treads in place and felt that an octagonal post would make it easier to shape them to fit. The post was a 10x10 Douglas fir timber, 26 ft. long, which I had to special-order. Once I got it inside my shop I laid it across two sawhorses and chamfered the corners with a Skilsaw. This required tacking a fence on the post and making an initial pass at the greatest depth of cut, and then making an additional rip on the adjacent side of the timber to cut through. To remove the saw marks and even out irregularities I used a belt sander—first across the grain and then lengthwise, trying to get the chamfered surface straight and uniformly wide.

Setting the post was easy, in spite of its length and weight. My son Greg and I slid the post down the hillside steps right up to the pier, and with a rope tied to the top, upended it until we could reach it while standing on the deck. With Greg steadying it from the top, I bear-hugged the bottom end, lifted it onto the pier and seated it over the ½-in. dia. steel pin protruding from the concrete. With the post roped to the deck rail, slightly out of position, I cut out a half-octagonal pocket into the deck band, slid the post into place and nailed it.

On top of the post I cut a pocket for the cross-beam of double 2x6s to carry the roof of the stair enclosure, then beveled off the top of the post for a neater appearance. The roof joists were set on a ledger on the wall and cantilevered over the beam (drawing, facing page). All the sheathing for roof and walls was 2x6 tongue-and-groove fir. Letting the roof deckings run long, I found the center of the post and swung a 3-ft. radius arc. To trim the decking to this line, I used a handsaw for the plumb cuts on the joists, and a reciprocating saw for the curve. I cleaned up the sawn edges by sanding to the scribed line with my belt sander. This new section of roof was finished with flat-seam terne metal held tightly under the soffit, and no changes were made on the existing roof.

Treads and risers—The new basement door was to be directly under the new living-room door, so I marked its location and punched a hole in the basement wall to find the floor level. With a 2-ft. level on top of a straightedge I transferred this elevation to the post. Next I measured the total rise on the post, and calculated the number of risers required to get a riser height between 7 in. and 8 in.—safe, comfortable limits for a single stair rise. I decided on 12 risers. To make sure I had divided correctly and converted the hundredths into sixteenths, I set my dividers and stepped up the post, making sure the top of the last riser would be exactly at the deck level. This required several attempts. An error of only ¹⁄₁₆ in. can make a difference of ¾ in. on twelve steps. When everything was right, I marked and numbered the location of each tread on the post. That done, I went back to the shop.

To lay out a stairway of this complexity, I like to draw full-size plans. On top of two work-benches, I laid a big sheet of clean cardboard

The completed stairway addition to this ranch-style house gives access to the basement without wasting interior space. The new enclosure also offers a covered entry at the deck level much nearer to street parking than was the previous front door.

and with trammel points, I swung a 3-ft. radius arc the width of the stairway. I had a cut-off from the post so I placed it right over the center and traced the octagon onto the cardboard. From the basement landing to the living-room deck, the stairs needed to spiral 180°, so I marked off a semicircle. Using the dividers, I stepped off 11 equal segments, the correct number of treads for 12 risers. Connecting these marks to the centerpoint gave the size and shape of each tread, as shown at right. These radius lines were actually the front face of each riser, and I drew in the line of the 1½-in. nosing on the tread for clarity. Now the pattern was drawn, and all I had to do was transfer the lines on the cardboard to wood. This is where the octagonal post is better than a circular one. You can make straight or angled cuts instead of rounded ones. I did no mortising and used no fancy joinery, but some of the intersections of the treads made excellent connections, almost locking themselves into place on the facets of the octagon. I rounded the outer ends of the treads with the reciprocating saw and belt-sanded to the line where necessary.

Finally, I grooved the underside of each tread to receive the top edge of the riser. Assembly went very well. Starting at the bottom, I put a little glue on the end of the first riser and toenailed it against the post with galvanized common nails, using a drift pin to set the heads below the surface of the wood. Galvanized nails were needed not for their rust resistance but for the coarse shank that would make them resist withdrawal as the treads flexed in use. Placing the first tread over the riser with plenty of glue in the groove, I nailed it securely. After the first two treads were in position and the glue had cured, they were strong enough for me to sit on, and as I added each step, I could walk up and down those already in place. With each new tread I also checked the tread heights against the layout lines.

The next thing was to cut the basement door in the block foundation. I used a masonry blade in my Skilsaw rather than knocking out the concrete block and laying new half-block in a sawtooth pattern. Cutting masonry with an abrasive blade creates clouds of dust and particles. You need an open work area and a good dust mask. This is why I made these cuts before enclosing the stairs. When the wall was cut through and the new door and jamb set, I bridged over to the post for the landing at the bottom of the stairs to form the entry to the basement. This bridge was supported at the block wall on a ledger, and by the post and a temporary prop to the ground on the stair side.

The enclosing curtain wall was also 2x6 fir decking. At the top, each piece was nailed to a joist or to blocking; the bottom was nailed to a plate on the deck, and as I progressed downward, it was nailed to the risers and treads. The boards were held flush at the top and allowed to extend long on the bottom. Again, I applied glue to all joints for strength. The tight joints between the steps and the post allowed no play in the stair, and the vertical fir sheathing connection to the roof added further rigidity and shear strength. The bridge into the basement

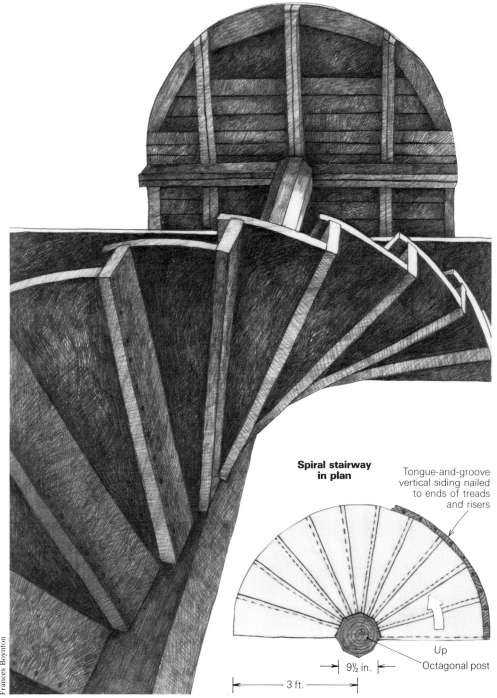

Frances Boynton

The spiraling pattern of treads and risers was laid out full scale in the shop. The bottom of each tread was plowed ½ in. deep to accept the top of the riser; the riser bottom was glued and face-nailed to the back of the tread. When the steps had been glued and nailed to the octagonal post, they were rigid enough to hold the weight of the builder, even without siding to support them. The roof for the stairs is carried by the double 2x6 beam let into the top of the post.

Labels in diagram: **Spiral stairway in plan** · Tongue-and-groove vertical siding nailed to ends of treads and risers · Up · 9½ in. · 3 ft. · Octagonal post

was suspended by the vertical sheathing, which was nailed to the deck joists above; then the temporary prop was kicked out. On the deck level, I cut a doorway into the shell of the stairway on the street side, with a large casement window across from it.

When all the siding was in place with the ends long at the bottom, I marked out the helix, transferring the bottom edge of each riser to the outside of the sheathing. The resulting series of points wrapped around and up the cylinder. Using an ⅛-in. strip of wood, I connected the points to create a line. Starting at the top to get the help of gravity, I just sawed down the length with the reciprocating saw. I took a lot of care with this cut so that it

wouldn't require any further dressing. The stairway was complete.

After cutting out the living-room wall and setting a new jamb and door, the only thing left to do was to remove the old stairs, cover the framed-in opening with oak flooring and finish it like the rest.

The work was done during October and November, and the delightful weather added to my enjoyment of the project. I had a helper for two days when I was mixing concrete and setting the post, but the rest of the time I worked alone. Including repairs to the inside floor and walls, the job took me eighteen days. □

Tom Law is a builder in Davidsonville, Md.

Newport Stairway Reproduction

Laminating stringers and handrails around a skeletal form

by Dale Nelson

In the fall of 1983, fire gutted one of the fine old mansions along Bellevue Avenue in Newport, R. I., completely destroying the second and third floors. The family that owned Oceanview had lived there for a long time. And because of their deep attachment to the place, they wanted to preserve whatever remained and rebuild the rest exactly as it had been. But once the salvage operation began, it was apparent that the original structure wasn't worth saving. They reluctantly decided to demolish the old building completely, then set out to reproduce the original as faithfully as possible, including the half-turn staircase.

The original oak staircase, built in 1885, rose three stories around a curved stairwell with a skylight above it. There were two intermediate landings between floors, and the open ends of the treads and risers curved as they approached and departed from these landings. The stairway was closed on the right but open around the stairwell, with a winding mahogany handrail supported by wrought-iron balusters and decorative scrolls.

What little remained of the original stair had settled in a pile at the bottom of the stairwell, covered by plaster and ash. The workmen sifting through the debris found many of the old balusters and scrolls, parts of the handrail, and even some large portions of the inside stringer. Everything was saved for future reference. Out of this debris and ash rose the new stair.

After the new house was framed and the exterior completed, my company won the bid to reproduce the staircase. The principal objective was to build the staircase so that all visible details were like the original. The owners encouraged us to repair and reuse as many of the salvaged parts as we could. As it turned out, the iron balusters were the only parts that we could resurrect. But the charred wooden remnants did provide clues to the original construction.

Since we had no detailed plans of how the original staircase was constructed, we carefully examined the remains of the stringers and studied old family photos showing details of the stairway. We discovered that the treads did not overhang the stringer on the open side, as is usually the case, but rather they butted into the stringer and were supported by blocks underneath—not a very rugged design. The risers, on the other hand, were mortised into the stringer about 1 in. back from the edge instead of being mitered into it. Cove molding was installed under the tread nosing.

The original curved stringer was very substantial; it was a bent lamination, in places twelve ⅛-in. plies thick. The balusters were composed of two ⅝-in. by ¾-in. iron posts connected at their centers by an iron ring. They were located in the stringer at every second tread, alternating with the decorative scrolls. The iron dovetails forged on the bottom of the balusters were screwed into neatly cut mortises that locked them in place. They could not have been cut with a chisel after the whole stringer had been laminated, nor could the screws have been put in afterwards since no screw holes or bungs were visible. The stringer must have been built in place and the balusters attached midway

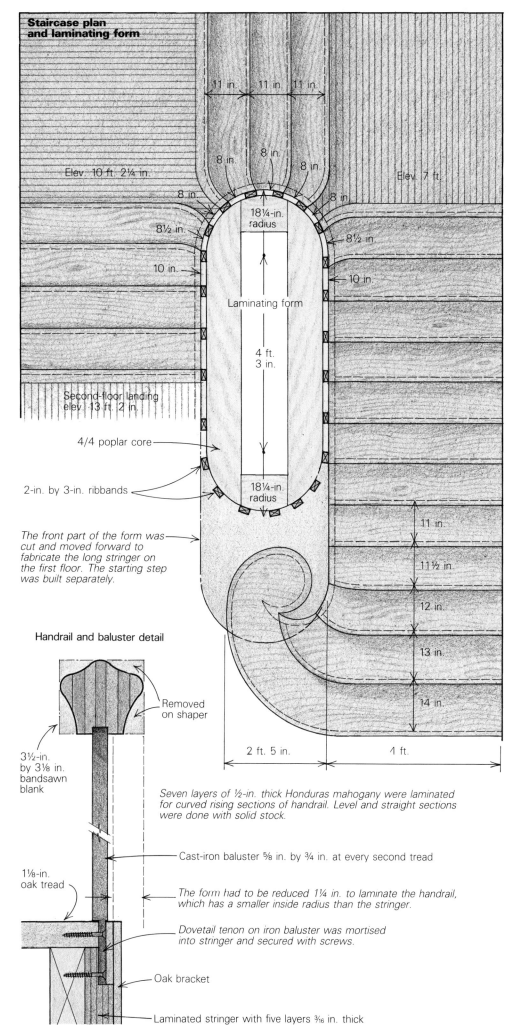

Staircase plan and laminating form

11 in.　11 in.　11 in.

8 in.　8 in.　8 in.

Elev. 10 ft. 2¼ in.

8 in.

8½ in.

18¼-in. radius

Elev. 7 ft.

8 in.

8½ in.

10 in.

Laminating form

10 in.

4 ft. 3 in.

Second-floor landing elev. 13 ft. 2 in.

4/4 poplar core

18¼-in. radius

2-in. by 3-in. ribbands

11 in.

11½ in.

12 in.

13 in.

14 in.

The front part of the form was cut and moved forward to fabricate the long stringer on the first floor. The starting step was built separately.

2 ft. 5 in.　1 ft.

Handrail and baluster detail

Removed on shaper

3½-in. by 3⅛ in. bandsawn blank

Seven layers of ½-in. thick Honduras mahogany were laminated for curved rising sections of handrail. Level and straight sections were done with solid stock.

Cast-iron baluster ⅝ in. by ¾ in. at every second tread

1⅛-in. oak tread

The form had to be reduced 1¼ in. to laminate the handrail, which has a smaller inside radius than the stringer.

Dovetail tenon on iron baluster was mortised into stringer and secured with screws.

Oak bracket

Laminated stringer with five layers ³⁄₁₆ in. thick

Nelson laminated the curved stringers around a form built in the shape of the stairwell, often gluing up two or three different stringers at one time. The ³⁄₁₆-in. oak plies were coated with glue, wrapped around the bending form, and clamped in place using oak cauls to distribute pressure across the lamination.

through the laminating process, with subsequent laminations covering the tenons of the balusters.

During construction of the new house, rough-framed stairs had been built to provide access to the second and third floors. Rather than demolish this rough stair, we decided to use it to support the new staircase. Therefore our stringers didn't need to be as massive as the original.

Layout and form building—The first step was measuring the elevations and dimensions of each rough landing, the overall dimensions of the stairwell, and the rise and run of each flight. After checking the measurements three times, we went back to the shop and laid out the staircase full scale on plywood, much as you would when lofting a boat. Like the measurements taken on site, this lofting had to be done accurately because it was the basis of all further work.

To laminate the inside stringers, we built a form in the shape of the stairwell, using our lofted drawing as a reference. The form was made of five horizontal 4/4 poplar cores (or templates) connected and braced by 2x3 vertical members called ribbands (photo above). The

ribbands were positioned around the form to coincide with the leading edge of each tread. But before attaching them, we placed the ribbands side by side on sawhorses, aligned their ends and marked the increments of rise for each step. When the rise and the run are uniform, even though curved, positioning the laminations is relatively straightforward, as long as the stock is wide enough. In our stair, the run was changing quickly as the stringers made the sharp turns at the ends of the well (drawing, previous page). We had to be extra careful that the layout did not "walk off the stringer." To accommodate these changes in run, we also had to make our laminating stock 14 in. wide.

The finished form was approximately 3 ft. wide, 7 ft. 4 in. long, and 10 ft. high (more height was unnecessary since the floor-to-floor height on the longest flight was only about 8 ft.). The form was leveled, plumbed, and then braced to the shop's roof purlins, deck and adjacent wall. We used 3-in. drywall screws to secure the form since it would be subjected to tremendous pressure from clamps and from wood in tension once the laminating process began.

Stock preparation and laminating—When the form was finished, the next job was to resaw the 5/4 red oak lumber into three equal plies. The 16-ft. long, 12-in. wide boards had to be fed slowly through our antique 36-in. Crescent bandsaw in order to keep the blade from wandering and ruining valuable stock.

Once resawing was completed, we thicknessplaned the stock to ³⁄₁₆ in. We carefully inspected each piece before sending it through the planer because oak this thin can disintegrate if the grain is irregular. There is always a strong chance of cupping with newly resawn boards, so we stacked the plies and clamped them with cauls until we were ready to use them.

At the same time that one man was preparing laminations, another was making treads, risers and trim pieces. The method of picking up patterns from the layout varied from person to person, but usually we laid the stock on the drawing and marked it directly. In some cases, where the stock was very large for instance, we used patterns to transfer the shape from the lofted drawing to the stock. The treads and risers were cut, numbered and stacked for later use.

To hold the lamination while the glue dried, we used 6/4 oak cauls about 16 in. long. and lots of clamps (photo above left). This laminating job was complicated by the fact that we needed a 14-in. wide stringer, when our stock was only 10 in. to 11 in. wide. Therefore, we had to edge-join the laminations at the same time we bent them around the form. We took one-third of the 10-in. wide plies and ripped them to 4 in., being sure we had two good edges. These pieces, combined with the 10-in. pieces, gave us the width required.

Next we laid the plies on a wide horse and picked out the pieces with the best color and figure. These were laminated first and last since they would be visible. Before applying the glue, all the cauls were laid nearby and the clamp jaws opened to the approximate gap.

We used Titebond yellow glue, applied with a

paint roller. Except for the first and last plies, both sides were coated, as well as the edges that were being joined. We had to move fast to keep the pieces from sticking together in the wrong position. We found that two plies, comprising four pieces, was the most two men could handle comfortably, especially where the clamping required a ladder. Also, trying to glue up too many plies at once increases the risk of springback. This can ruin the accuracy of the radius. Gluing up only a few layers at a time and allowing the glue to cure eliminates this problem.

The plies were brought carefully to the form and temporarily clamped in place with large handscrews. Starting in the center and working in both directions, we then clamped the cauls to the ribbands, being careful that the edge-joined pieces butted into each other and didn't accidentally slip and overlap, creating a bulge in the lamination. We staggered these joints, starting one time with a 10-in. piece and the next time with a 4-in. piece, so that they didn't all line up and weaken the stringer. Short pipe clamps were applied across the lamination to make sure the edges of the plies were tight. Gluing up is always a nerve-racking process, and this was especially true with these stringers. The work required our undivided attention; telephone calls went unanswered.

Routing for treads and risers—When the laminated stringer was finished, we cut it to rough length with a circular saw, and began transferring the tread and riser information from the lofted drawing to the stringer. Once again, we worked with the awareness that a mistake in our layout would have disastrous consequences, since an erroneous router cut couldn't be concealed. We used triangular pieces of plastic laminate as flexible pitch blocks to step off the rise and run on the stringer.

Next we routed the mortises for the treads and risers, using templates clamped to the stringer. The ones we use are similar to conventional stair templates, but they're made of 6mm Baltic birch plywood that's kerfed so that it will bend to conform to the curve of the stringer.

It's important to size the cutout in the template so the fit is snug but not so tight that assembly is difficult. We routed the housings for the treads and risers with a double-flute, carbide-tipped ³⁄₄-in. stair bit, guided around the template by a fixed bushing on the router. The stair bit resembles a dovetail bit but has less angle. It undercuts the housing slightly, ensuring a proper fit at the edges. Since approximately ½ in. of stock must be removed in one pass, we used a 3½-hp heavy-duty router.

When we were routing the stringers, we were also working on the curved risers. The radius of these curves varied according to the riser's position on the staircase (drawing, previous page). For these risers we used ³⁄₄-in. red oak lumber-core plywood and kerfed the back of the curved areas (photo facing page, right). Once the risers were bent to the proper shape, we filled the saw kerfs with epoxy and then glued on a face of oak veneer over this area to strengthen it.

When the curved and straight stringers had both been routed, the pieces of the staircase

Nelson set up adjacent flights in his shop (left), positioning them at their correct heights relative to each other in order to fit the joint between their stringers. For the curved risers, Nelson used red oak lumber-core plywood with saw kerfs. But the tight scroll of the starting step (above) demanded a combination of kerfed plywood for the first part of the curve and veneer for the rest.

were assembled quite easily. We screwed the treads to the curved stringer, positioning the screws so that they would be hidden by the decorative trim brackets on the outside of the stringer. The treads and risers were glued and screwed to each other and glue blocks applied every 6 in.

After assembling the two flights leading to the third floor, we set them up in the shop at their correct heights relative to each other (photo above left). This preliminary setup allowed us to check the alignment and to fair the areas of intersection on the stringers. We then made curved gusset strips out of ½-in. plywood that we screwed to the backside of the stringers to hold the two flights together.

When all pertinent measurements were checked, the uppermost flight was taken down and set aside for delivery to the job site. The lower flight was raised to the higher position and the next flight was assembled and brought up underneath, so that the fairing and checking could continue. This process was repeated with all the flights except the larger one on the first floor, for which we had to raise the adjoining flight a bit higher.

The actual installation of the various flights at the job site was reasonably straightforward. Each was placed over the rough stringers, shimmed and screwed in position. When each flight was secured, the lower one was brought up, the gussets fastened, and the flight attached to the rough stringers and plastered walls.

Handrailing—The sections of railing fell into four categories: the straight pieces, the curved level pieces, the curved rising pieces, and the easings, which form a transition from a rising section to a level section. We ground two sets of shaper knives, one for the side and one for the top of the railing, taking the patterns from an old piece of rail that had survived the fire. We always use lock-edge steel for our knives, and we make sure the knives are sharp and well balanced. Once the knives were set, we ran the straight sections of handrail, using 16/4 Honduras mahogany stock.

The flat curved sections were laid out on solid stock with patterns taken from the lofted drawing, glued up and bandsawn to rough shape. Then we ran them on the shaper, using one jig to shape the inner curve and another one for the outer curve.

When these were done, we started on the curved rising portions of railing, which were the most difficult part of the job. The changing tread runs at the ends of the stairwell were part of the problem. These meant that the pitch of the railing would be greater where the run was shorter and less where the run was longer. Keeping an equal distance from the handrail to the leading edge of the tread would cause the railing to kink and destroy its graceful contours. We created a curve that looked right, and accepted the fact that the distance to the leading edge of the treads would vary through this area.

We glued up bent laminations for these sections of the handrail because segmenting solid stock would mean running some end-grain sections through the shaper, and this can be dangerous. The form had to be modified since the railing was thicker than the stringer and therefore had a smaller inside diameter (drawing, p. 61). We ripped down the ribbands to account for this extra thickness, making sure we retained our riser layout when we reassembled the form. Then we ripped the mahogany stock into ½-in. by 4-in. strips and laminated them in the same way we did the stringers.

When we finished the laminating, we squared up the edges of the rail on the bandsaw, using a curved fixture clamped as a fence to the band-saw table. This was essential because we needed a uniform cross section from which to gauge the profile of our molding work. Since these pieces were too awkward to run all the way through on the shaper, we also used a combination of the router, various gouges, rasps and spokeshaves. Sometimes, where a single piece of railing rises and twists quickly, a machine setup just isn't practical or safe, and using hand tools actually speeds the job along.

While the final woodworking was being done on the handrail, we sorted out the old balusters and decorative scrolls, which had been sand-blasted and painted. This took a long time because the lengths, widths and shapes were different. The balusters were cut at various angles depending on their location in the original stair. Some of the scrolls were curved, some were flat. It turned out that we needed four new balusters, but fortunately the scrolls, whose intricate iron work would have been very hard to reproduce, were all there.

Since the staircase has no intermediate newel posts, the balusters are the main support for the handrail. We mortised the dovetail tenons into the stringer through the outside face and secured them with screws. These mortises were then covered by the scalloped brackets that we cut out on the bandsaw from ⅜-in. oak. The tops of the balusters as well as both ends of the decorative scrolls were simply screwed in place.

All sections of the railing were cut to length on site, fitted with rail bolts and dowels for alignment, and installed piece by piece. We then faired them with hand tools and sandpaper.

Building and installing the stairway took approximately 475 man-hours and was done over a four-week period. □

Dale Nelson is an architectural woodworker in Newport, R. I. He specializes in staircases.

Storage Stair

An alternative to conventional framing takes advantage of normally wasted space

by Tom Bender

A stairway that my wife Lane and I saw in Japan about 10 years ago etched itself in our memories. Instead of being built on notched stringers, it was made up of finely joined chests and boxes of graduated heights. In stepping from the top of one box to the top of another, you arrived at the second floor. Needless to say, when the time came to build a stair in our own house recently, the Japanese stairway was still in our minds. Our stair (photo, left) contains four different kinds of storage units: drawers, cupboards, a roll-out toybox and a book alcove. There's even a drawer under the lowest tread that pulls out for storing slippers and socks. Considering all the stuff that we're able to store in this normally wasted space, building the storage stair wasn't that much work. It was more work, of course, than for a conventional stairway, but no more than for a storage wall.

Design and layout—Our visions of fine Japanese joinery had to adjust to our own reality. We had an old Sears table saw, a skillsaw, a cheap belt sander, and a random assortment of hand tools. Our cabinet shop was a corner of the living room with the rug rolled up. We couldn't afford good hardwood, and didn't want a house too precious for people to live in. So we had to make do with the pile of construction-grade fir stacked in our living room, some used shiplap, and someone who had never built a stair before.

The structural system I worked out for the stair is a framework supported by four vertical partitions that serve as dividers between drawers or cabinets. These are connected by a grid of 2-in. by 1½-in. face-frame pieces that form a base for treads and act as drawer and cabinet bottoms.

I built all the partitions first. Each one is a different height, but they are all constructed the same way. An identical pair of 2x4 uprights was grooved to accept a ½-in. thick plywood panel and notched at the top for the 2x10 riser that is an integral part of each partition. I drew up a checklist of all the framing members and hung it up by the table saw, then crosscut long stock to rough length with my skillsaw and brought it to the table saw for finish cutting. The 2x4s for each partition were cut in pairs, since they would run clear

Tom Bender is an architect and builder. He lives in Nehalem, Ore. Photo by the author.

to the floor. I chiseled shallow (⅛-in.) mortises in the 2x4s to accept 1½-in. by 2-in. horizontal face-frame pieces that would connect adjoining partitions and also support the shelves, drawers and treads. These pieces would be joined to the 2x4s with glue and dowels, so I made a cardboard template and stuck a nail through it to mark dowel centers.

Assembly—Putting the whole thing together would have been easy for an octopus with a truckload of clamps. But I was armed with only two bar clamps, a few C-clamps and two hands. Partitions were assembled first. While they dried, I located their positions on the floor, cut all the horizontal pieces to size and checked dowel alignment on each joint. For the final assembly, I put up the tallest partition first, at the high end of the stairs. After gluing and doweling all eight of the horizontal pieces to this partition, I snugged the unit into place with the aid of my faithful bar clamps. After checking for level, I toenailed it to the floor. The remaining three partitions went up in the same way. Each one was clamped to its steady, previously installed neighbor before being nailed to the floor.

We let the glue in the completed framework dry overnight, and began on the stair's ten treads. As all our clamps would be needed to hold a single tread in position until the glue dried, I decided to nail-clamp the joints to hold them while the glue set. I drilled undersized holes, then nailed through the back of the upper riser with 16d box nails, and up through the side frame with 10d finishing nails. Then the bar and C-clamps could be removed and used to set the next tread. After a touch-up sanding, the frame and treads were finished with two coats of boiled linseed oil.

Doors, drawers and shelves—With a usable stairway complete, installing the shelves, drawer glides and stops, and building the doors and drawers could go more slowly. To match the walls of the living room and entry, I made all the doors with the same recycled shiplap, using horizontal battens on the inside face of each door to hold the shiplap boards together. Drawers were made up from ½-in. plywood, with Masonite bottoms. Shiplap was glued up to make each drawer face. For the handrail, I used a long piece of driftwood that we had found on the beach. I cut it to length, then notched it to fit against a 3x3 post that I attached to the stair frame with lag screws.

A translucent screen—Next, we needed to enclose the living-room side of the stairs to make the stairway safe for kids and also to keep heat downstairs. We didn't want the stair and entry to be dark, so a translucent wall seemed like a good idea. Everything pointed toward a Japanese-type screen. The modified *shoji* design we used was easy and inexpensive to build.

A frame for the translucent wall was made from a few 2x4s ripped into strips. I installed ¾-in. by 1½-in. strips vertically, attaching them to the edges of the treads and to the

Varied storage, step by step

Connecting four vertical partitions with 1½-in. by 2-in. crosspieces creates a stepped face-frame that supports the stair treads and also provides storage compartments. Each partition spans the width of the stairs and has 2x4 sides that support a let-in 2x10 riser and horizontal plywood panels.

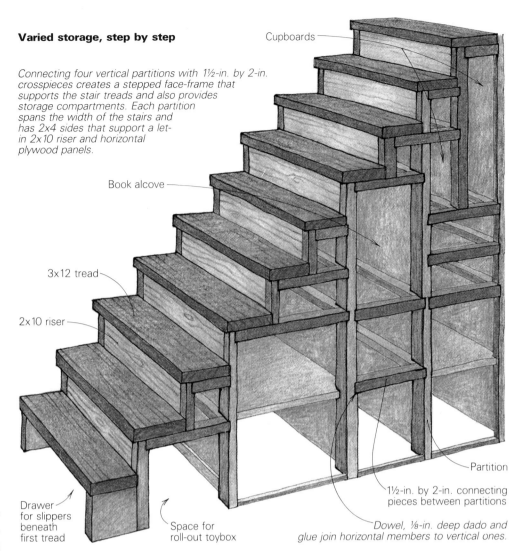

Cupboards

Book alcove

3x12 tread

2x10 riser

Drawer for slippers beneath first tread

Space for roll-out toybox

Partition

1½-in. by 2-in. connecting pieces between partitions

Dowel, ⅛-in. deep dado and glue join horizontal members to vertical ones.

edge of the floor above, which made them strong enough to prevent anyone from falling through. Smaller ⅜-in. by 1-in. strips were used horizontally, spaced half a riser apart, to stiffen the frame and support the paper covering. I notched all the pieces to a depth of ½ in. at all joints, gang-cutting with a dado blade on my circular saw to ensure correct alignment.

After a trial assembly to make sure everything lined up properly, the notches were then dabbed with glue, and the strips snugged into place with the aid of C-clamps. The inevitable curvature of the long, thin strips of wood caused some concern at first, but any unevenness was pulled back into line as the grid grew. The frame was sealed with a coat of linseed oil, except on the surfaces where the paper would be attached, and allowed to dry well, so that it wouldn't bleed onto the rice paper that would be glued onto it.

To apply the rice paper, I used a paste made of white flour and water. This makes it easy to remove the paper if any sections get damaged. The rice paper, which is commonly used for *sumi* ink drawing, is available in 8½-in. and 11-in. wide rolls from most art or oriental-import stores. We used the larger size, applying it vertically in sheets no longer than about 2½ ft. We found that if we used longer lengths, the glue we painted on the lattice framework would dry before the paper was in place. We worked from one end of the wall to the other, doing one vertical section at a time. A little

thumb pressure stretched out any wrinkles or sags, and we were able to keep the paper taut against the frame without much trouble. After waiting a few minutes for the glue to dry, I trimmed off the excess paper with a single-edge razor blade. Then we applied the next strip of paper, lapping the first by the thickness of the wood frame.

A door with rice-paper panels at the stairway finished off the job. Double-swinging spring hinges on the door allow us to go in and out with the ever-present armfuls of children, groceries, laundry or firewood, without losing too much heat or having to stop and close the door behind us. A wide mid-rail on the door acts as a bumper bar, allowing us to open the door even with our arms full.

The finished stair and storage space have met almost all our expectations. The rice paper creates a soft light that accentuates the rich color of the steps. And at night, an almost magic silhouette of the gnarled, curving handrail is cast against the precise and delicate rectangles of the grid.

We had feared that the rice paper wouldn't last long with small children around, but we have all learned to live with its delicacy pretty well. Occasionally holes do get punched through it, but they are easily patched with a rice-paper snowflake or butterfly. The storage works out beautifully, with sweaters, blankets and such available from either the entry or the living-room side of the stair. ☐

Staircase Renovation

Loose, tilted flights signal trouble underfoot, but even major problems can be fixed with screws, blocks and braces

by Joseph Kitchel

There seems to be a great deal of reluctance among renovators to tackle any kind of stair rebuilding. I have seen many beautifully renovated homes with sagging staircases, loose balusters and makeshift railing supports. People who wouldn't think of leaving cracked plaster cracked or sagging floors unshored will put up with staircases so crooked they would make a sailor seasick, and squeaks so loud they wake the whole family at night.

Procrastination and lack of understanding of the mechanics of stair construction are at fault here. No one seems to know quite where or how to start, or who to call to do the job. To be sure, attacking an ailing stair takes courage, resolution and perseverance. It is one of the dirtiest and most disruptive renovating jobs.

The stairway is the spine of the multistoried row house. It links the sometimes minimal area of individual floors into what can be a spacious and accommodating floor plan. Because changing the stairway can seriously affect the physical flow of people and spoil the aesthetics of the whole interior design, it is far better to renovate than redesign or reposition.

Parts of a stair—Each step consists of a riser, the vertical portion that determines the height of each step, and the tread, the part on which you step. The dimensions of steps must be consistent within flights, though they may vary from one flight to the next. Variation within the flight will break the stair-user's physical rhythm and cause tripping. Tread widths may vary in the bottom two or three steps of the main flight for aesthetic purposes, and at the top of a flight, where pie-shaped steps are needed to negotiate a curve. Riser height, however, must not vary.

The newel post, usually decorated with paneling or carving, supports the railing at the bottom of the flight. Though it gives the impression of being heavy and sturdy, after years of being swung around by ebullient children, it has probably loosened enough to sway from side to side or lean out toward the hall. The vertical supports of the railing, aligned more or less behind the newel post and marching upward with each suc-

cessive step, are balusters or spindles, which are dovetailed into the treads.

Keeping the balusters from slipping out of their joints on the open side of the staircase are the noses, continuations of the molding that forms the front edge of the treads. These noses are removable and not part of the tread itself.

On each side of the stairs are the stringers, which hold the risers and treads. The stringer attached to the wall is usually routed or plowed out to receive the steps; this is called a housed stringer, and it produces a strong, dust-tight stairway. The outside stringer may be open or closed. Stringers may be simple, laminated with other pieces to form the curves of the stairs, or partially concealed by decorative filigree.

Were you to remove the plaster under the staircase, you might find two or three 4x4s or 4x6s running the length of the flight—one nailed to the inside of the outside stringer, one centered beneath the steps, and perhaps an-

other nailed to the inside of the stringer attached to the wall. These are the carriages, the members that add extra support. At the bottom, the carriages ideally rest on top of the stairway header, a joist that frames the stairwell end. They are sometimes attached to the inside face of the header. (This was the cause of failure in one flight I repaired. The weight of the stair forced out the toenails holding the carriage.) At the top, the carriages attach to the face of the upper-story header joist or, in the case of a curved flight, to angled braces running from notches in the wall to the upper floor joists.

Other elements of stair construction visible from beneath the flight are the wedges or shims driven between steps and stringers. Tapping these wedges tight or adding wedges made from shingles or building shims can do a great deal toward tightening up a staircase.

Diagnosis and dismantling—Problems with stairs fall into two categories. The simpler ones concern the railing, balusters, newel post and railing supports—the superstructure. Trouble here, though relatively easy to repair, can be symptomatic of more serious problems in the steps, carriages and stringers—the substructure.

Before you remove any plaster, explore the failings of the flight. If the steps are loose and tip toward one side, then the carriage or stringer on that side has weakened. The cause may be rotting, breaking, warping and splitting, or the carriage may have separated from the stringer.

Another symptom of major deterioration is a series of gaps or cracks along the ends of the treads or risers where they fit into the stringer. These rifts may occur on either side of the stairs, but are most often on the wall side, suggesting that a center or outside carriage has shifted downward, skewing the flight toward the center of the stairwell.

Large cracks in the plaster at the top or bottom of a flight are a good sign that the carriages have come loose from the headers. However, cracks generally parallel to the carriages or crisscrossing their length may indicate that vibration has caused the plaster to loosen and the keys to

From *Fine Homebuilding* magazine (February 1981) 1:14-18

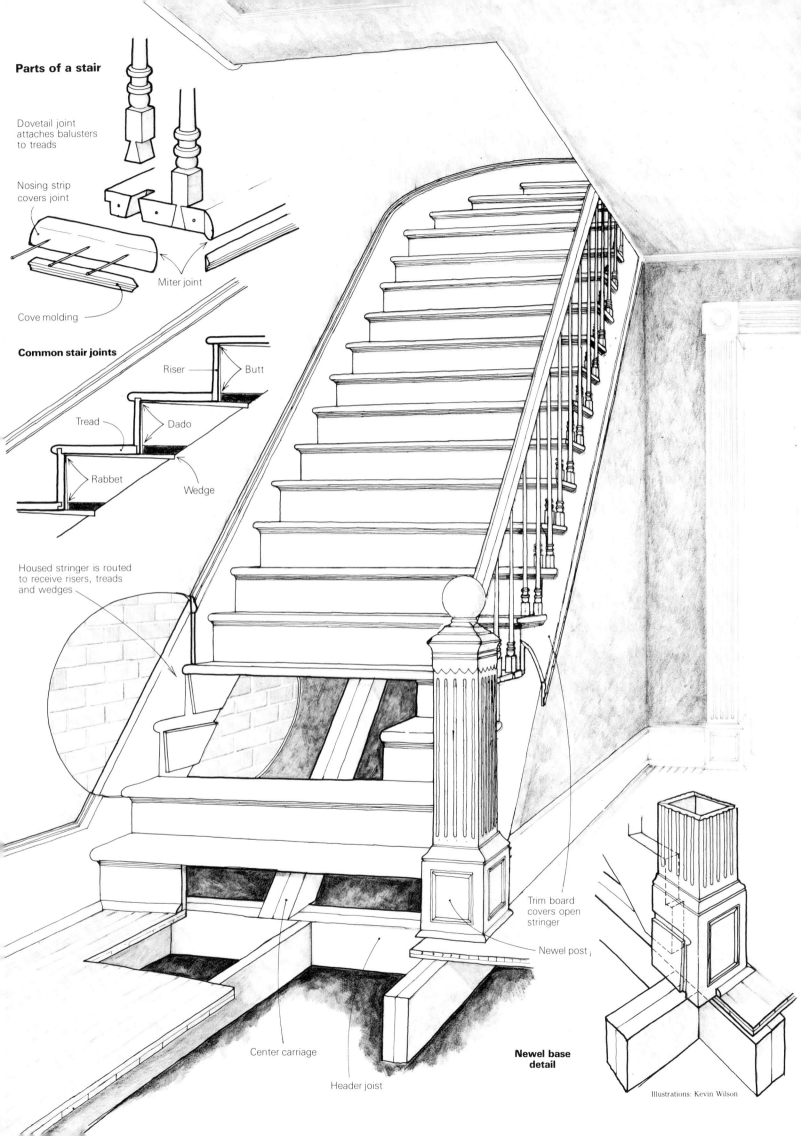

Parts of a stair

Dovetail joint attaches balusters to treads

Nosing strip covers joint

Miter joint

Cove molding

Common stair joints

Riser

Butt

Tread

Dado

Rabbet

Wedge

Housed stringer is routed to receive risers, treads and wedges

Trim board covers open stringer

Newel post

Center carriage

Header joist

Newel base detail

Illustrations: Kevin Wilson

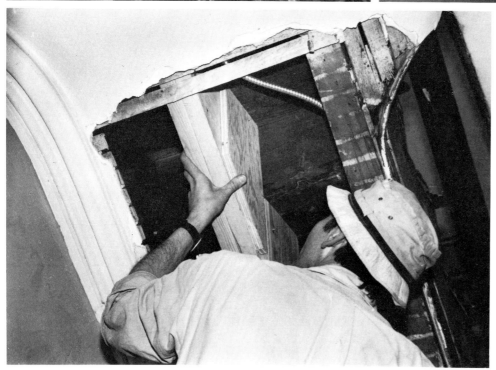

Substructure repairs are major problems and should be tackled first. To correct a distorted stringer, author Kitchel, above left, strips stairs of treads and risers. Plaster has been removed from beneath the stairway previously, and the resulting debris cleared away for a dust-free work area. The jack between wall and stringer provides lateral support until the stringer has been repositioned and reattached, and all steps have been renailed. Above, a carriage pulled away from the joist can be reattached with an angle iron made from $\frac{1}{8}$-in. plate attached to the header with $\frac{3}{8}$-in. bolts. The carriage usually rests on top of the stairway header, but occasionally it is attached to the inside face. Left, support blocks of $\frac{3}{4}$-in. plywood fastened with Sheetrock screws minimize deflection.

spindles, may have been purely cosmetic. Remove them so that all parts of the stairs may be properly aligned.

Substructure repair—You should now know the causes of your stair failure; tackle the bigger ones first. If a carriage is rotted out or cracked, tear it out and replace it. If it is intact but springy, it may be undersized; laminate a new beam or a steel reinforcing plate onto it. Solutions at this point must be as individual as the problems. But by far the most common failings are bolts and nails which have worked loose, allowing carrying elements to pull free from walls and header. In this case, mending plates made from $\frac{1}{8}$-in. steel work well.

If you must raise or remove carriages, first free them from each riser and tread; otherwise, the attached superstructure will loosen as the skewed members are jacked up. To remove treads and risers, pry them apart at joints or cut through the nail shanks if pulling is impractical.

For major repairs, jack the stringers or carriages into place in a manner that won't damage the superstructure and will keep it from damaging you. To give the jack or brace a level bearing surface, attach angle blocks cut to the slope of the stairs with clamps or screws.

break off, which is a much less serious repair.

If your stair has a decorative plaster molding running along its wall side, a continuation of the ceiling cornice, don't despair. This can be saved if it is not badly damaged and is still attached securely to the wall. To dismantle, carefully cut through the plaster and lath along the molding with a masonry blade in a circular saw, cutting parallel to the stair edge of the molding, leaving the molding intact. Plan this cut so that plaster or Sheetrock can be rejoined to this edge.

To further assess causes of stair failure, remove nearby plaster and lath. You may not have to remove all of the ceiling covering if the plaster isn't bad and if the problem is localized. Take time to clean up all the resulting plaster dirt. Debris allowed to accumulate on the flight below makes moving your stair platform difficult and dangerous, and will also worsen the

spread of dust throughout the rest of the house.

If necessary, remove noses and balusters, but label all parts first. Assign each step a number (I usually start at the bottom) and tape this number to the outside of the step and to its adjacent nose molding and spindle.

Using a screwdriver or chisel, gently pry the noses away from the stringer. They're usually finish-nailed in two or three places. You'll see the dovetail joint that connects the spindle to the tread. Tap the spindles out at the bottom and pull them down out of the railing.

Examine the joint between riser and tread to judge how to handle repair of squeaks or gaps. The joint may be a dado, a rabbet or simply butted and nailed. Past repairs, such as wedges driven into the gaps between riser and tread, fillers in the spaces above the treads along the wall stringers, or braces along the railing or

If the wall stringer has come loose from a masonry wall, reattach it by raising it to its proper position and nailing it with cut-steel masonry nails of sufficient length to go well into the wall. Wear goggles. A better method is drilling through the stringer and into the masonry with a carbide bit, and tapping in a lead sleeve; a lag bolt and washer expand the sleeve and tighten the stringer to the wall. If it's a frame wall, lag screws alone will hold stringers to studs.

If the outside stringer has twisted or warped, and is pulling the treads out of the wall stringer opposite, remove the treads and risers and force the stringer in toward the wall. Get the necessary leverage by temporarily bracing from the outside of the stringer toward the partition wall opposite. Screwing or bolting the stringer to the accompanying carriage will correct matters.

With the supporting members repositioned, now attend to the steps. Repair split treads and risers by removing them, lapping a piece of plywood over the back of the split, and gluing and clamping overnight. If you glue and screw the lapping piece you may forego clamping, and can replace the piece immediately. If the very edge of the nose is split, insert dowels from the edge to reattach it, being careful not to split the riser's dado or rabbet joint.

After you've corrected stringer and carriage problems and each tread and riser is back in place and renailed to the outside stringer, strengthen the steps by nailing step blocks to the center carriage. Why this wasn't done originally has always puzzled me. From a piece of ¾-in. plywood cut a step block to fit under each tread, and place it firmly against the back of each riser. Nail or screw the blocks to the side of the carriage, and then nail through the face of the tread and riser into the edge of the blocks. Trim the bottom edges of the blocks to conform to the angle of the carriage. I usually alternate the blocks on opposite sides of the carriage, but they all may be attached to the same side if the staircase is narrow and you can't get between the center carriage and the wall. Nailing blocks on both sides of the carriage is overkill, but add them wherever extra support is needed.

At this time a test run up and down the stairs will tell you where additional nailing and bracing are needed. For all face nailing in treads and risers, I use 6d or 8d finish-head, spiral flooring nails. For nailing where it doesn't show, I prefer 6d or 8d rosin or cement-coated box nails. I find 1½-in. or 2-in. Sheetrock screws driven with a variable-speed drill useful where hammer space is limited. Screws often add more strength than nails, because they pull things together and don't require pounding, which may disturb the alignment of nearby areas.

If risers, treads or stringers are to be refinished separately from the spindles, consider doing this now. Stripping, sanding and painting are easily done with the upper parts out of the way. Before finish is applied, set and fill all nails.

Superstructure repair—Newel-post problems are best dealt with after all other structural problems are solved. Although removing or loosening the newel post may be necessary to work on the carriages or stringers, it can usually be left in place to support the railing.

Newels are usually attached to the bottom step with a threaded rod, and to the railing with a hanger bolt. (Hanger bolts have wood-screw threads on one end and machine-screw threads on the other.) On the machine-screw thread of the hanger bolt is a star nut, which can be turned through the access hole with a screwdriver or needlenose pliers after the bolt is in place. (A plug fills the access hole later.) For extra strength screw into the railing from the inside of the newel post.

Straighten or tighten a shaky newel post by shimming under its bottom edges and renailing or screwing it to the floor. For greater support, try one of these repairs using a threaded rod. First, remove the newel post. Bolt a threaded rod to a bracket or wooden block so that the bolt fits flush with the bottom of the block. Screw the block and rod assembly to the floor, then slip the newel post over them and reattach it. To repair the post without removing it, attach two brackets under the tread of the first step as

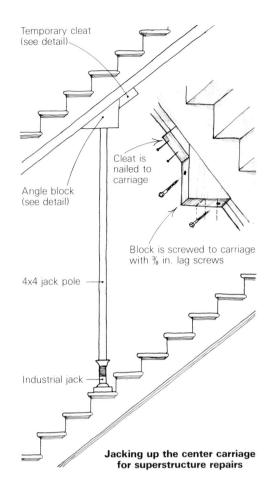

Temporary cleat (see detail)

Cleat is nailed to carriage

Angle block (see detail)

Block is screwed to carriage with ⅜ in. lag screws

4x4 jack pole

Industrial jack

Jacking up the center carriage for superstructure repairs

Building a Stair Platform

Before you renovate your staircase, you'll need to build a stair platform. Part of the reason stair repair is so difficult is that there is no place to stand, no way to get up to the job.

The exact dimensions of the platform depend upon your stairs. To make the platform, set a straightedge, level, on the fourth or fifth step from the bottom of your main staircase. The height of the platform is the distance from that step to the floor; its length is the distance from the back of that step to the front of the bottom step. Check these dimensions at different locations on various flights of your staircase. Sometimes rise and tread dimensions or angles of incline vary from flight to flight, but not usually. Then cut and attach legs and braces as shown in the drawing.

The platform should be wide enough to hold your stepladder comfortably, but narrow enough to allow passage on the stairs when it's in position. The platform will probably fit your neighbor's stairs, and you can use it when repainting your own stairwell; it is therefore a tool to retain after renovaton. You might consider making it collapsible for easier storage.   —J.M.K.

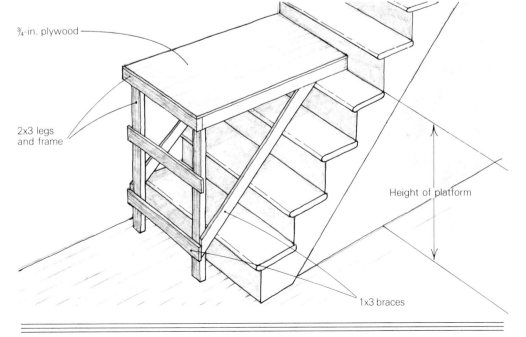

¾-in. plywood

2x3 legs and frame

Height of platform

1x3 braces

Newel-post attachment

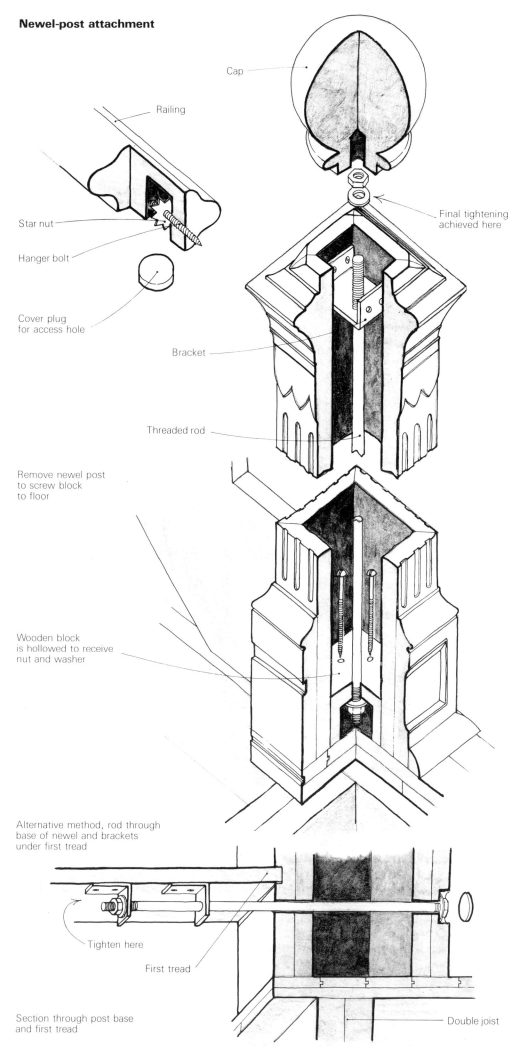

Cap

Railing

Star nut

Hanger bolt

Cover plug
for access hole

Final tightening
achieved here

Bracket

Threaded rod

Remove newel post
to screw block
to floor

Wooden block
is hollowed to receive
nut and washer

Alternative method, rod through
base of newel and brackets
under first tread

Tighten here

First tread

Section through post base
and first tread

Double joist

shown at left. Drill a hole through the base of the newel post to align with the holes in the brackets. Insert a threaded rod through the holes, bolt at both ends, and fill the access hole with a plug.

To replace the spindles, first coat both ends with white glue. Insert the top end of the spindle into the railing underside first, then slide the dovetailed end into its tread slot. Shim the joint wherever necessary, from underneath or from either edge, securing it with a finishing nail through the dovetail into the end of the tread. I find a rubber mallet useful when replacing spindles, because it will not mar the finished surface of the wood.

As you proceed from bottom to top, occasionally check the alignment of the railing and make adjustments by trimming or lengthening the spindles. Temporary braces hold the railing in place until all spindles have been installed and the glue is dry.

You can lengthen spindles (or adapt spindles from another stair) by adding a short piece of dowel. Drill into the top end of the spindle, and glue and nail the piece of dowel in. When the glue is dry, rasp and sand the dowel to the contour and taper of the spindle tip. Stain to complete the match.

When the final spindle has been inserted, the small bracket that originally connected the top of the rail should be reattached, or a new one made to fit. This prevents lateral movement of the rail and will hold the spindles in their correct positions as the glue dries.

Refinish the noses before you attach them. If you remove old nails by pulling them through from the back with a pair of nippers, you'll avoid the splitting that usually occurs when the nails are pounded back through and pulled from the face. Glue the noses and nail them twice along the side and once through the miter where nose meets tread molding.

Replacing the ceiling under the flight is the final step; use lath and plaster or Sheetrock. If your stairs curve, use short sections of wire lath to recreate the original curve.

Take care to keep all nailing surfaces in the same plane. This can be done with two straightedges; one the width of the ceiling area, from the outside of the stringer to the wall, and the other as long as possible to run the length of the flight. Use building shims where necessary to keep furring strips in the correct plane. Determine your plaster line in relation to any plaster molding along the wall and to the bottom edge of trim pieces that adorn the outer carriage. I usually let the beaded or molded edge of this outer trim protrude below the plaster line. Existing pieces of plaster or plaster molding to be replaced may be drilled and secured with screws before touchup spackling or painting.   □

Joe Kitchel, 42, is a prop builder, cabinetmaker and renovator from Brooklyn, N.Y.

A Revamped Victorian
Builder rescues a Queen Anne cottage in distress

by Barry Schwartz

Michael Zelver is a builder, and so he knows what kind of chaos is generated by a full-scale renovation project. Consequently, when he and his wife Ann Wasserman began looking for a run-down Victorian house to remodel, they vowed that they would finish the project before moving in. They knew that if they lived in the house during construction, the job might never get finished.

Zelver says, "I have seen a number of people buy behemoth houses to renovate, and they just never finish. Someone will go out and buy a wonderful Victorian waiting for restoration. It's 5,000 sq. ft. and it's sitting right on the ground; it's really more than one person can handle. I

was looking for a house that I could renovate by myself. What I had was lots of time and very little money."

Finding the house—After a search of about six months, Zelver and Wasserman found what they were looking for—a 1,000-sq. ft. Queen Anne style cottage built in 1890, in Santa Cruz, Calif. The house still had its architectural integrity, but it was nearly derelict. Transients had been living in it, and an unmistakable pall of decay had settled over the house. Vines were creeping into its gutters, many of its windows were broken, and in one place you had to circumnavigate a hole in the floor to avoid falling

into the crawl space. Even Zelver's builder friends were dubious about renovating the structure. But the house sat on a good-size lot, and it was in a decent part of town. The couple closed the deal, and started thinking about changes.

First Zelver completely gutted the inside of the house, removing all non-structural partition walls and ceilings. Then he jacked up the entire building, in preparation for a concrete foundation. The original perimeter footings were brick, and posts from the central girder bore on blocks of wood resting on the ground.

The house was built into a slight slope, and its crawl space had from 3 ft. to 5 ft. of head room. Because he knew he would have to do extensive

Nearly derelict after years of neglect, this Victorian cottage, left, has been completely refurbished by its new owners, Michael Zelver and Ann Wasserman. Much of the new exterior construction was concentrated at the bay window (above). Here Zelver replaced the original siding and trim work from the window header down, and increased the venting area with more latticework directly above the block foundation. Trim caps made of closet poles conceal gaps between the sash trim on both sides of the large bay window.

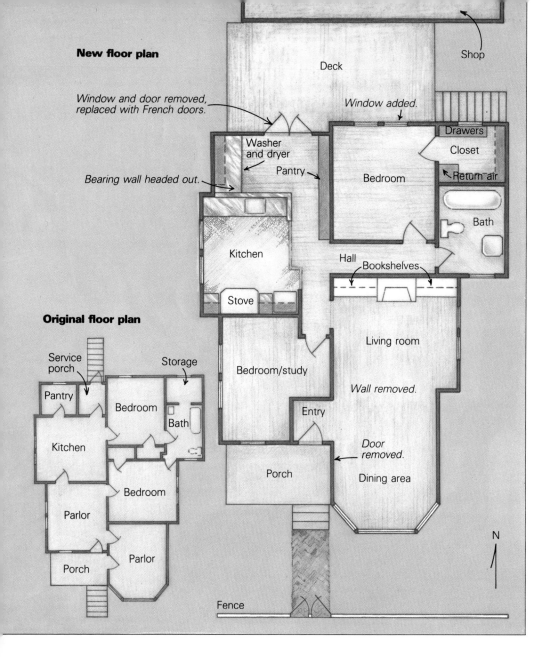

New floor plan

Shop

Deck

Window and door removed,
replaced with French doors.

Window added.

Washer
and dryer

Drawers

Closet

Pantry

Bedroom

Return air

Bearing wall headed out.

Bath

Kitchen

Hall

Bookshelves

Stove

Original floor plan

Service
porch

Storage

Pantry

Bedroom

Bedroom/study

Living room

Bath

Wall removed.

Kitchen

Entry

Bedroom

Door
removed.

Parlor

Porch

Dining area

Parlor

Porch

N

Fence

structural repairs to the floor framing and re-build it mechanically, Zelver wanted to retain the luxury of all that room. After the house was lifted up, he built a perimeter block foundation and set the house back down on pony walls.

About a third of the joists were badly termite-eaten, and rather than simply sistering the weak joists, Zelver decided to remove them, along with any threat of continued infestation. The floor above consisted of 1x T&G fir, blind-nailed at an angle through the tongue. To remove the joists without damaging the flooring, he cut them completely through 1 ft. back from the sill with a Sawzall, and cranked them over so that he could see the angle of the nails coming down from above. He then beat the joists out in the appropriate direction, pried out the two small end pieces, snipped the nails off flush, and installed new joists. Originally the house had only a center girder supporting 2x10s on a 14-ft. span but that proved to be too springy, so he added another girder line down each side.

Changing the floor plan—Zelver could now concentrate on the interior layout, and it took four designs before he got it right. Like many Victorians, the original house (drawing, above left) had lots of little rooms. Zelver remembers,

"The floor plan was a disaster. You had to walk through a room to get to any other room. No hall, no central plan whatever."

Zelver's goal was to create a more open plan with larger, light-filled rooms while still retaining the original Victorian character of the house. But he couldn't do it arbitrarily because there were offsets all over the house and the windows were all very low to the ground—a typical feature of Victorian homes. Zelver's solution (new floor plan, above) was to remove one of the two entries, merge one of the parlors and one of the bedrooms into a combination living room/dining room, and to transform a pair of closets into a hallway to the master bedroom and the bathroom. At the back of the house, where there had been three small rooms, he placed a kitchen and floor-to-ceiling pantry cabinets, including a washer and dryer hidden beneath a built-in hutch. Another bedroom/study sits at the juncture of the kitchen and the living room.

Bringing back the details—When Zelver demolished the interior, he kept samples of the original baseboard and fancy trim from the living room. He planned to reuse what he could, and mill whatever he needed to finish the trim work. Since he planned to paint the trim, Zelver

bought the least expensive material he could find that is suitable for decorative millwork. This turned out to be roughsawn, 4/4 molding-grade pine stock, clear and flat-grained. Zelver surfaced the stock with pattern-ground knives, mounted in a 12-in. jointer.

Zelver estimates it would have cost about $3,000 if he had had a shop do the work, but as it was, the whole procedure took him two days, and cost $500 for the material.

The medallions with the concentric circles at the corners of the door and window casings are called rosettes, and Zelver needed quite a few of them. He found a local woodworker who is adept at lathe work and was willing to spin off copies for $2 apiece. Zelver's job was to supply him with the blanks. Like the molding, the medallion blanks are paint-grade pine. This detail is consistent throughout the house, except in the entry to the new study. There the passageway is too narrow to allow full-width casings around all the doors. Rather than abandon the rosette detail, Zelver bandsawed radiused blocks for the concentric-circle carvings. The resulting detail (photo facing page, bottom left) maintains the continuity in a tight space.

Plaster walls and moldings—Because the house was to be plastered rather than taped and textured, the finish trim and the cabinets were all set first. The plaster comes right up to their edges. In an old out-of-square house, there are advantages to this system. For instance, if a corner doesn't meet at exactly 90°, it doesn't create a problem with baseboards. They are simply laid up and shimmed to a tight fit. The plaster makes up for any imperfections by meeting the baseboard rather than the other way around. Another advantage to this process is the setting of electrical fixtures into the wet plaster. This eliminates unsightly gaps and air leaks.

The crew Zelver hired did the walls and ceilings, as well as the curved baseboard around the fireplace and the crown molding. About half the cost of the plastering went into the ceiling crown molding. Out of a ten-man crew, four people worked on it exclusively. The molding is a duplicate of the ceiling molding in the Capitol Rotunda in Sacramento, Calif. The same crew had done a major restoration there, and they still had the metal forms used to shape the plaster moldings into an even profile. They thought the design used at the Capitol Building would look nice on the living-room ceiling. Zelver and Wasserman agreed.

After the crew installed the gypsum lath, they snapped a horizontal line on the wall just below the crown molding. Then they attached a temporary ledger board to act as a guide. The plaster molding is built up a little at a time, and it adheres directly to the wall with no other support. As the plaster is hocked up in small amounts, a worker uses a tin form backed with wood to shape the wet plaster. The form registers on the ledger.

In the first five passes it doesn't look like much, but by the sixth or seventh pass, the molding begins to take shape (photo facing page, top right). The plaster is hardening the entire time, and the plasterers can't stop until it's

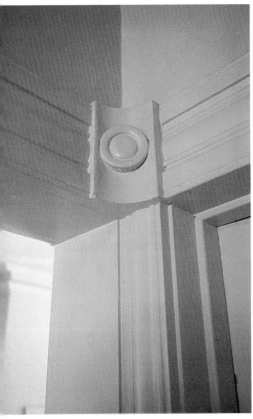

Plaster moldings. Two plasterers work a crown-molding form along a ledger affixed to the wall (top right). The form has metal edges, and wood backing for strength. After repeated passes, the detail begins to take shape. This process works only for straight runs. With a plaster crown molding, both inside and outside corners have to be tooled by hand. The concave central core of this pattern is especially attractive at an outside corner (top left). The completed living room (bottom right) has a dining area on the south side, and an entry to the west. The fireplace is a new addition. It is faced with marble the owners found at a salvage yard, and flanked by built-in bookcases. Patterns of light cast by the leaded-glass window in the bay play across the floor. All of the trim, including the baseboards, rosettes and casings around the door, matches the original trim. It's tough to fit two rosettes into a tight corner, but a radiused version (above) allows just that, without a loss of continuity in the trim work.

Construction photos: Ann Wasserman

A grain hopper (above) has found a new life as a stove hood in the cottage's kitchen. A custom-made rectangular flue carries off the cooking vapors to a roof-mounted fan. In the laundry room (below) a washer and dryer are concealed behind raised-panel style cabinet doors. Note the thin layer of plywood on the back of the door. The inset panel is glued to it. The wall cabinet above uses the same detailing, and it's capped with a piece of pine crown molding. The cabinets are finished with satin Man-O-War spar varnish.

Zelver brought more light into the kitchen by replacing a bearing wall with a dropped header. That let him install a skylight and a pair of French doors to the back deck. Non-structural plaster brackets at each end of the dropped header help to harmonize the beam's presence in its Victorian surroundings. The kitchen floor is hexagonal tile—the rest of the flooring is #3 red oak shorts.

done, not even for lunch. Occasionally an entire section will simply fall off, and the process has to begin again.

By the last pass, the molding is nearly complete and the plaster is fairly dry and solid. But not completely. On the very last pass one plasterer pushes the form along the molding so hard that it squeaks while another one wearing rubber gloves pushes gobs of plaster into the back of the form so that a very thin film sets onto the molding, filling in any knicks or cracks. The result is a surface so smooth that it doesn't even require sanding.

The job, however, has only begun at this point. The plasterers are only able to maneuver the mold within a foot of all the corners, both inside and outside. Corners have to be built up

gradually by hand and eye, without benefit of the tin form.

The primary tool used for this job is called a *mitering rod*. It's a piece of flat steel with a 45° degree angle at its end. Once the straight runs of molding are completed, the rod is used as a guide against the hardened plaster. Little by little, small amounts of plaster are again hocked up onto the ragged edge of the molding, and with small trowels, slowly built up toward the mitered corner. Outside miters are easier, inside miters are the toughest. It takes a practiced hand about two hours to do either one, and the results (photo previous page, top left) are worth it. The curved baseboard next to the fireplace was formed in much the same way, with a form cut to the contour of the baseboard stock.

The kitchen and the laundry—The original kitchen had little natural light, and four doorways breaking up the walls. Consequently, there wasn't a logical place for a continuous expanse of countertop. To get the doorway count under control, Zelver removed the bearing wall between the kitchen and the two tiny rooms to the north—the pantry and the service porch. The combined pantry/porch is now the laundry room. He made up for the structural change with a dropped header that is embellished with elaborate plaster brackets (photo above left).

To add more natural light to the combined space, Zelver installed a pair of French doors that open onto the back deck and a large skylight in the roof over the laundry area. The ceiling over the laundry follows the pitch of the

From *Fine Homebuilding* magazine (August 1985) 28:49-53

roof, which eliminates the need for a deep sky-light well. Zelver points out that deep skylight wells look entirely out of place in a Victorian house. He downplayed the skylight's aluminum frame by masking it with a wood frame on the inside. In addition, its mullions carefully match and align with those in the French doors. To further the continuity between the skylight and the doors, they have the same casings and rosettes.

With the old wall out of the way, Zelver could add a counter peninsula that forms a low wall between the laundry and the kitchen (photo facing page, left). The low windows on the west wall were a factor in the placement of the counter, and the window casing and cabinets are carefully integrated at their junction.

The peninsula countertop is salvaged marble. Before its transformation it was wainscoting in a Santa Cruz bank. Based on the marble sheets he bought, Zelver drew up a detailed blueprint showing where he needed cutlines, beveled edges, outlet holes and the sink cutout. He had the work done by a company that specializes in stone monuments.

Another piece of salvage that looks right at home in the kitchen is the hood over the 1918 stove. It's a small grain hopper, installed upside down and painted to match the stove (photo facing page, top right). Zelver had a sheet-metal shop fabricate a rectangular flue to fit atop the hood. A roof-mounted fan provides the suction.

The laundry area is behind the counter, and the washer and dryer are housed by cabinets consistent with those in the kitchen. Like the casings around the windows and doors, the cabinets are made of flat-grained, molding-grade pine. While he likes the look of raised panel doors, Zelver wanted to avoid the extra expense of slotting the rails and stiles, and making floating panels. Instead, he had his cabinet-maker dowel together face frames, then glue 1/8-in. lauan plywood across the back of the frames (photo facing page, bottom right). The plywood came from dinged doorskins, scrounged from a door company for $2 apiece. The plywood was cut a bit oversize and allowed to run wild across the frames. Then it was trimmed flush with a bearing-under trim bit. The panels were then cut to the inside dimension of the frames, shaped with a raised-panel bit on a shaper, and glued to the plywood.

Rebuilding the bay—Even though the new foundation was straight and level, the building had settled unequally during its first 90 years. Framing members had torqued slightly, making it impossible to get smoothly mitered intersections in some of the trim work without substantial reconstruction or tedious fiddling. One such junction occurs between the vertical trim framing the bay windows. Rather than caulk the gaps, paint it, and be done with it, Zelver added a clever detail to conceal the gaps. He used two lengths of 1⅜-in. dia. closet pole, as caps (photo p. 71, right). He cut a lengthwise 90° notch out of the backs of the poles, so that the sharp edges of the cuts dig into the window trim, ensuring a tight fit. Starting at the top of the junction, Zelver nailed the caps in place. As he worked down, he torqued each cap slightly to

match the eccentric window trim. In this manner, the inaccuracies are spread out, and become invisible in the finished exterior.

Victorian fence—The front of the house is bordered by Zelver's version of a Victorian-style fence (photo right). Unlike most wood fences, this one looks the same from front and back, like wrought iron. The rails are made up of three pieces, which together equal the 3-in. dimension of the posts (drawing, right). The outside portions of the rails are continuous, while the interior members are 4 in. long. The pickets are sandwiched by the outside rails, and evenly spaced by the interior segments. Lengths of half-rounds decorate the sides of the rails and extend across the posts, providing a continuous detail. The tops of the rails are ripped to a 30° angle, providing both a Victorian detail and a surface that sheds water. The posts are topped with caps beveled at 15°.

Zelver wanted sharp edges throughout, so he remilled the construction-grade redwood down to a slightly smaller dimension. And he had another reason for reducing the size of the stock. "Architecturally I don't care for 3½-in. by 1½-in. material. It's too clumsy. Also I wanted a nice crisp edge everywhere, and everything you buy has a bullnosed edge."

He spent a day setting the posts, and another one milling the stock. Then he put the entire 40-ft. fence together in four hours.

"And the whole thing," Zelver says, "is predicated on the nail gun. If you had to hand-nail all these individual pieces, not only would it be time-consuming, but it would also be very difficult to hold them properly while you nail them. It could be done, but it would be tedious to say the least. This fence was extremely satisfying to build, because, after all the milling, you go outside and put it together, and it is instant gratification. The design allows it to be somewhat forgiving, and it all tended to shrink together uniformly. If the initial execution is carefully done, then as the fence degrades after a while, it still looks good."

Zelver believes in the power of paint. Like any good designer or builder, he thinks past the present moment to how the finished product will look, whether it is a fence or an entire house.

"Everybody starts out as a wood butcher and likes wood. But there is something that happens to things architecturally when you paint them that doesn't exist when it's natural wood. A number of people who watched me working on the house couldn't understand what was going on, and when eventually it all got painted they said, 'My God, I had never realized how important paint is before.'

"When I look at the work in progress I'm thinking of people who don't focus on architecture at all. All they get is an impression of what you did, and it's either a good impression or a bad one. They are not going to pick up on the details, they are not going to know why you were successful. All they are going to know is that they like or don't like what they see." □

Barry Schwartz is a contractor and writer. He lives in Santa Cruz, Calif.

The Victorian-style fence is a wooden version of a wrought-iron fence. It looks the same from both sides, as shown in the plan and section drawings below.

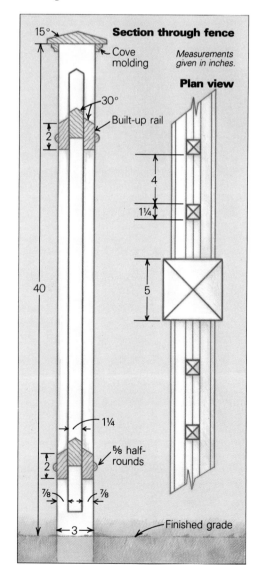

Section through fence

Cove molding

Measurements given in inches.

Plan view

15°

30°

Built-up rail

2

4

1¼

40

5

1¼

⅝ half-rounds

2

⅞ ⅞

3

Finished grade

Ornamental Plaster in England

Elegant designs from hide glue, powdered chalk and classical antiquity

by Donald E. Wahlberg

In Great Britain the definition of a fine home often presupposes the use of traditional materials and finishes, such as handmade brick, fine woodwork and plastered walls and ceilings. These can include cornices, paneling and foliated ornament. One consequence of this prevailing taste in domestic architecture is that the admirable tradition of decorative plasterwork developed in the 18th century has been carried on ever since.

Decorative features in plaster, things like cornices and architraves, derive from classical design, and these composed the basic vocabulary of the decorative arts in 18th-century Europe. The influence of Italian arts and crafts upon European artistic thinking was considerable at this time, and it was incumbent upon any serious British artist to tour Italy and examine the works of ancient Rome, of the Renaissance and the Baroque period. This peripatetic education stimulated the desire to use and extend the language of classical architecture.

Robert Adam and Kenwood House—The brothers James and Robert Adam were England's leading architects during the latter half of the 18th century, and Robert Adam created a recognizably personal style of decoration, which received notable patronage. An excellent example of his mature style can be found at Kenwood House in London. Situated at Hampstead Heath, the house has long been associated with Scottish affairs in Great Britain. It was acquired in 1754 by Scotsman William Murray from a fellow Scot, the third Earl of Bute, who was a close adviser to George III, and later Prime Minister.

Murray became the first Earl of Mansfield and was a leading light of his time as well as an architectural patron, serving as Attorney General and Lord Chief Justice. He was influential in law reforms and gave a number of controversial decisions in favor of civil rights in cases concerning slavery and religious freedom.

After 10 years at Kenwood, Lord Mansfield chose fellow Scot Robert Adam to remodel the house. Adam was nearing the height of his career, having returned from Italy in 1758, where he was greatly influenced by antique Roman decoration. He was at a point in his development as a designer where his own style was emerging.

The house Adam had to work with dated from the late 17th century and was a largely brick two-story home with an attached orangery. Given unrestricted freedom in the design, Adam added a third story to the main block and a new library wing to balance the orangery, which itself was remodeled to match the overall composition. Later additions and refurbishments to the house were made by other architects, but much of Adam's work can still be seen, partly as a result of a restoration in 1975. The house has been a public picture gallery since 1928.

The entrance hall is a fine example of Adam's style and shows many of the effects that can be achieved with decorative plasterwork. The harmony of the decorative elements—ceiling, cornice frieze, mantelpiece, architrave and wainscoting—illustrates the effectiveness of achieving a correct balance of these in a room. The ceiling decoration is most delicate, and is well suited to

the scale of the room. In reference to this ceiling, Sir John Summerson, a renowned authority on classical design, has commented that "in the execution it has slightly more elegance than in the drawing," which reminds us of the importance of artistry and craftsmanship to the success of a design.

One of the best of Adam's rooms is the library, which was also used as a reception room (photo facing page). It is a much stronger and grander space than the entrance hall, and its arched ceiling was an innovation in England at the time, with its apsidal ends framed by pairs of columns. The handling of the plaster enrichment here is well balanced to the color scheme and the furnishings. The plaster ornaments, executed by Joseph Rose, include a cornice with lions and heads of deer, which are references to Lord Mansfield's coat of arms. The painted panels are by Antonio Zucchi, who had traveled and worked with Adam in Italy.

The marble hall and music anteroom are the work of a later architect, George Saunders, and show the excellent use of plaster enrichments to complement the structure and form of these rooms, and Adam's work in adjacent rooms.

The library at Kenwood House, facing page, is one of the most successful rooms designed by the 18th-century architect Robert Adam. The gilded plaster frieze and column capitals, as well as the foliated filigree on the ceiling, were cast in a London shop before being installed. As a plastic material, plaster came into its own in the second half of the 18th century, chiefly because new recipes made it possible to cast large sections of cornice and ceiling panels without cracks. Today delicate plasterwork like the ceiling ornament above is done using composition plaster, a mix of glue, linseed oil, ground resin, water and ground chalk.

Early plastering techniques—The decorative features used at Kenwood House are derived from classical motifs that were typically expressed in stone and wood carvings. Executing such detailing in these media required highly skilled and labor-intensive work, and even in the 18th century, the cost of producing it was prohibitive for a private individual. In order to satisfy the demand for decorative plasterwork, substitute materials were sought after to reduce production time and costs. And so composition plaster came to the rescue.

Robert Adam purchased a plaster recipe concocted by a Swiss pastor. Generically known as *carton pierre*, the mixture was composed of a binding material—usually animal glue—and powdered chalk. This fine-grained plaster allowed very delicate designs to be cast numerous times in the same mold. Adam passed this recipe on to his collaborator George Jackson, who later established a firm for producing decorative plaster. In the following century one of Jackson's sons bought the rights to a newly invented French process for producing fibrous plaster, which permitted large reinforced sections of decorative work to be fabricated in the shop and later fixed in place at the building site.

The firm of George Jackson and Sons is still in business today, and is located on the Thames, several miles upriver from the Houses of Parliament. The workshop is housed in a brick warehouse filled from basement to roofline with molds, including 30,000 original brass and boxwood molds from centuries past. The many styles available include Asian and Arabic motifs that have been produced in recent years to supply international demand.

At Jackson's, the techniques of production have been refined over the years, yet in its essentials the work has changed little. Although it is now a division of a larger corporation, the little company continues to operate as a craft-oriented enterprise, employing experienced artisans and taking on apprentices. As a result, the company enjoys its reputation as a firm that specializes in plasterwork, and has been responsible for the creation and renovation of many of the finest rooms and houses in Great Britain, including Kenwood House.

Two contemporary methods—The techniques used by present-day producers of decorative work are composition molding and fibrous plaster casting. Composition work, like the ceiling ornament shown in the photo at left, produces delicate designs, which are often applied to fine cabinet and carpentry work and are overpainted to appear as an integral part of the wood. Ornamental fireplace fronts and mantelpieces are produced in this manner, as are mirror and picture frames, thereby avoiding the cost of carving in stone or wood, while achieving superb sculptural form and finish.

A typical recipe for "London compo" calls for 18 lb. of Scotch (animal) glue, 2 qt. of linseed oil and 9 lb. of ground natural resin. The glue is dissolved in 3 qt. of water and stirred continuously. In a separate container, the linseed oil and resin are heated together, dissolved and stirred. The sized water is fully mixed with the

From *Fine Homebuilding* magazine (December 1987) 43:40-43
Photos, except where noted: Donald E. Wahlberg

Compo plaster ornament, like the wall molding above right, is cast in a rigid mold (above left), which is withdrawn before the mixture fully cures. After the piece is trimmed to final size with a knife, it's allowed to set up. To apply the ornament, the back of the piece is softened by steaming, and it is stuck on pins projecting from the surface it's meant to cover. In the photo below, workers are cleaning a vinyl cornice mold before brushing the first coat of plaster into every tiny recess.

oil and resin, and finely ground chalk is added to provide the consistency of a thin dough.

The compo is then kneaded on a chalk-dusted board, cut into strips and laid on the mold while still flexible. The mold has been lightly oiled with linseed. A damp board is then placed over the compo, which is mechanically pressed to fill every detail of the mold, and to make the compo adhere to the board. This allows for easy removal of the compo after molding (photo top left), and provides a base for trimming with a sharp knife. Trimming at match lines must be perfect so that filling of joints is not required.

Fibrous plaster casting is used for much larger pieces, like ceiling panels and cornices, and these are reinforced with plaster and wood, as will be explained further on.

Making molds—In the past, reverse molds for composition plaster were often made of metal, sulfur or handcarved boxwood, which comes from the evergreen box shrub, a species much favored by woodturners and engravers for its durability and its excellent workability (it takes and holds details crisply). These days, hard plaster or fiberglass are now more often used, along with plastic and latex molds.

The molds used to produce fibrous plaster are made by several methods. The problem with reusable molds is that they have to be removed intact once the plaster has hardened. This means that they have to have what foundrymen call "draft"—the slight taper that permits the removal of the molded object. This taper from bottom to top puts constraints on the design, and in the old days when molds were made of rigid materials, certain areas, called undercuts, had to be carved by hand or formed separately in elaborate piece molds.

But a mold that could be made to stretch around undercut areas when being removed would have serious advantages, and in the mid-19th century the gelatin mold was introduced. Good as it was, it suffered from a low melting temperature, and was adversely affected by the heat generated by setting plaster. Its softness required great care to avoid damaging the molded face, and these molds generally had short lives. After World War II, polyvinyl chloride (PVC) was introduced and came to be the most widely used mold material.

Modern molding techniques involve both hot and cold-pour methods, using polyvinyl chloride and polysulfide respectively, which are available in different degrees of flexibility.

Vinyl is a thermoplastic material that must be melted at temperatures of 215°F to 310°F, and the temperature must be carefully controlled because the compound will burn easily. The advantage of hot-pouring a mold is that vinyl will produce exactly the surface of the model, and it can be remelted and used again. In addition, seasoning of the mold (making it non-sticky) is less critical because the vinyl is oily and will not stick easily to the model. Apart from reproducing a fine degree of detail, it will accept both rough treatment and the high temperatures encountered in casting with some materials.

The material used for cold-molding is a polysulfide compound in liquid form, which sets when mixed with a curing agent to form a flexible rubber. The curing time can be adjusted. The liquid must be thoroughly mixed and poured slowly to ensure complete curing and to eliminate bubbles. Cold pouring does not have the drawback of hot-pour material, which can damage the surface of the model by its high heat. However, rubber molds do suffer from cold flow, and must be stored over the original models to retain their shape for successive uses. If properly stored, rubber molds will produce very fine detail time after time. When casting, the mold must be brushed with a releasing agent to keep it from sticking to the cured plaster.

The construction of a model and preparation for taking the mold are lengthy and detailed processes, as anyone with experience casting sculpture in art class will know. The important aspects are the correct construction of dams and joints, the sealing of porous surfaces and the reinforcement of the model. This is a painstaking process with either original work or reproduction molding.

The materials used for models are diverse and each has its own characteristics, but a firm that specializes in reproducing architectural elements will generally be taking molds from existing decoration or from wood, clay or plaster models.

Casting—Once made, the reverse mold is supported in a wood frame or cradle of wood and plaster construction. After it has been carefully cleaned (bottom photo at left), it's ready to be filled with plaster and reinforcement.

Fibrous casting is done with the two-gauge method. Plaster of Paris for the first and second coats is mixed in separate containers, and each has a different setting time. The first coat forms the face of the casting and has to be worked into the detail of the mold with brush and fingers, and is built up to a thickness of ¼ in. As this begins to set up, the first layer of reinforcing

Installation. After the plaster has been allowed to cure, it is ready to install. In the photo above, a section of a fibrous-plaster cornice is being mitered with an ordinary handsaw. The cut need not be very precise, as the joint is purposely left open (right). The gap will be stuffed with plaster-soaked fabric. When this cures, more plaster is added and then tooled flush (below). The finished miter is shown below right.

fabric—a rough-weave cloth made from jute yarn and called hessian—is stretched over the cast and pressed into the plaster. This layer of hessian is cut wider than the mold to allow it to be turned back over the wood laths at the edge of the casting. Then the second coat of plaster is brushed onto the hessian so that the two coats will bond together.

To reinforce such large pieces of plaster and to provide a means of attaching the ornament to a wall or ceiling, wood lath is added to the wet plaster at this point. The strips of wood are well soaked in water before being added to the plaster so they won't absorb moisture from the plaster of Paris (used here for whiteness) and swell during the curing process. The laths are bedded in plaster and wrapped in hessian that has been impregnated with plaster. A second layer of hessian, indispensable for a satisfactory result, is then applied in a similar manner to the first. The back of the casting may be finished with a light

splatter texture to reinforce the cast between laths without adding lots of weight.

A properly completed casting can be removed from the mold before the surplus plaster left in the mixing container has fully set. This is made possible by the correct gauging of materials in the mixes and the expert working of the plaster in the mold, both of which influence setting times during the casting operation.

Installation—For small ornaments, such as the wall design shown in the top right photo on the facing page, the compo is attached by softening the back of the piece with steam and fixing it in place with panel pins—tiny finishing nails that are driven carefully through the compo into the wood or plaster backing to hold it in place until the adhesive sets.

Fibrous plaster is generally used for larger work, including cornice sections, which are cast in lengths of 8 ft. to 10 ft. and typically rein-

forced with hessian and lath. The miter cuts for cornices are made with a handsaw (photo top left), and only two marks guide the cut. The cornice can be attached to the wall in several ways—with plugged screws, wire or with nails. But always the fastening happens through the embedded lath in the plaster. On especially heavy pieces, metal reinforcement is cast into the cornice to give it overall strength and to offer firm anchoring.

Once in place, the mitered joints, intentionally left open, are filled with hessian that's been soaked with wet plaster. After the wadding in the joint cures, the surface is built up and tooled flush with the rest of the profile (photo above left). Sometimes mitered corners are covered with a miter leaf, a cast ornament designed to hide any imperfections in the joint. □

Donald Wahlberg, a transplanted American, practices architecture in London.

Molding and Casting Materials

Plaster is still the old master, but rubber by-products give more options

by John Todaro

In restoration work there are times when the only way to reproduce an object is to mold it and pull a cast. This is the case with plaster moldings, ceiling rosettes, cornices, decorative plaster work, friezework, wood carvings and missing metalwork on furniture. With the proper materials it's possible to make a mold of any object: I've even molded high-relief patterned wallpaper. Why go to all the trouble? Usually, the parts are either impossible to get or are too expensive to have made professionally.

In making a mold or negative image of a part to be reproduced, your materials selection depends on factors such as cost, complexity of the part and the number of times you'll use the mold. In casting reproductions from molds, your choice of materials is affected by the surface finish wanted, weight, strength and other factors that vary with each application. The materials list I offer here is by no means definitive, but it should give you a good idea of the range of the products available to the restorer.

Molding materials

Plaster of Paris—Plaster, the all-time classic molding and casting material, is made from gypsum rock that has been heated to drive off the water. The resulting soft rock is crushed into a powder; when water is added, the powder recrystallizes into a solid as hard as the original gypsum rock.

Plaster is cheap, easy to use and doesn't shrink—important where dimensional accuracy is crucial—but large molds and castings can become quite heavy. Although plaster won't stick to the object you're molding or to the castings if you use a mold-release agent, casting objects with severe undercuts may require knockdown molds of several pieces. Because plaster has a short working life, professionals slow setup time by adding lime: generally one-third plaster to two-thirds lime for plastering walls. A teaspoon or so of liquid hide glue or fish glue added to the plaster will slow setup time by about two hours. Opened plaster absorbs moisture from the air and will form lumps when you try to mix it, so don't buy more than you need, and store what

John Todaro, 42, lives in Brooklyn, N.Y. He uses molding and casting techniques to restore ornamental plasterwork in old brownstones.

Ceiling rosette cast in plaster of Paris. (Photo by Jeff Fox.)

you don't use in airtight containers or tightly sealed bags in a dry place.

Moulage—A special molding material commonly used to take impressions from life (or, as the case may be, death), moulage should be used when an object is delicate or when its patina must be preserved. Napoleon's death mask was made using moulage. It captures the finest detail and, best of all, is reusable.

Moulage is usually used to create one master mold from which a more durable mold is made. To do this, heat a block of moulage until it melts. Carefully paint the moulage (while still warm) on the object to be molded, and allow it to cool and set. Before peeling the mold from the object, make a mother mold of plaster or, if the mold is large, from damp sand. This will provide support for the somewhat fragile moulage mold. Make the master cast (plaster is most commonly used) and from the cast make the master mold, using whatever material is most suitable for your purposes. To reuse the moulage, just reheat.

Liquid rubber latex—This material easily can become the restorer's workhorse; it is an excellent molding product that creates light, flexible, durable molds suitable for a variety of casting compounds. It's easy to use, sets quickly, reproduces detail well, doesn't shrink, has inherent release ability (though you may need to use a mold-release agent on complex parts), is tear and crack-resistant, and can be used to make molds of objects of any material. However, if the patina on the object to be molded must be preserved, don't use liquid rubber latex: This material, like plaster, can destroy an object's finish.

Liquid rubber latex comes with a filler; the more you add, the more rigid your mold will be. Since the filler is less expensive than the latex, using more filler will also extend your latex supply. This is a technique developed for mass production where costs are to be kept minimal, but this technique has its limits. I've found that one part latex to three parts filler is an acceptable proportion for molds used to mass produce objects with no undercuts; more filler than this will tend to make the material crumbly. If you are molding a piece with severe undercuts, however, reduce the amount of filler to provide the most flexible mold possible. It's a good idea to experiment. To capture maximum detail on intricate parts you can use the latex without filler for the first coat or two of a mold built up to about five coats, but don't construct the entire mold out of fillerless latex because it will be so flexible that it will lack "memory," and it won't snap back into the shape of the object you have just molded.

RTV silicone rubber—You can make molds of objects of almost any material with RTV (room-temperature-vulcanizing silicone rubber) including low-melt metals, and it is stiff enough so that a mother mold is not needed. It has inherent release ability, but on complex parts it's necessary to use a mold release or graphite. This gentle material won't destroy an object's patina, the way liquid rubber latex and plaster can. It captures finer detail than latex and is stronger, but it is more expensive.

RTV is a two-part medium: a liquid silicone-rubber base and a catalyst or curing agent. Once mixed, it sets quickly and hardens within an

From *Fine Homebuilding* magazine (February 1981) 1:36-39

hour. It is generally too thick to be brushed on, so you must use a box or a casting jig.

When using RTV silicone rubber, make sure your work area is adequately ventilated, and avoid breathing in fumes. The material may also cause skin irritation, so wear rubber gloves and wash off any compound that you happen to spill. These simple precautions also apply to a few of the other molding and casting materials mentioned here: polyester resin, Monzini compounds, polyurethane foam, and epoxy casting resins. It's essential to read manufacturer's directions and cautions carefully before using any molding or casting material.

Polyester resin—By alternating coats of polyester resin and layers of fiberglass you can build up rigid molds good for hundreds of casts. The advantages of this method are many: You can capture fine detail, molds are lightweight and are excellent for pliable castings, such as from latex rubber. They also work well for thin sheet castings or for large castings with light materials, such as polyurethane foam. This method is also used where RTV would not be economical, or where plaster would be inappropriate or too heavy. However, polyester resin molds are inflexible, and to cast undercuts, you will have to build a knockdown mold of several pieces. Careful preparation of the original with a release agent is imperative.

Polyester resin is used with a catalyst—once mixed it will set up within half an hour, so never mix more than you need. To thicken the polyester resin, add Cab-O-Sil (fumed silica, which comes in the form of light flakes) when mixing the resin and catalyst; it adds bulk and strength to the mold. How much you add depends on how thick you want the material to be. Ideally, it should pour slowly, like heavy honey—use too much Cab-O-Sil and the material won't pour, use too little and it will be runny.

Pour the resin/catalyst/silica mixture on the object, which has been treated with a release agent, and let it set. When the first coat has hardened, apply a layer of fiberglass. Paint or pour more polyester resin on the fiberglass, smooth with a piece of wood or a scraper, and let it set.

If you're making a large sheet mold, a few 2x4s or 2x2s or pieces of pipe will keep it from warping. Lay the braces on the back of the mold, then cover them with polyester resin and strips of fiberglass so that wood will bind to mold. This will also give you something to grip when holding the mold besides the mold's sharp edges. When fiberglass is used with polyester resin, you'll know why they call it fiberglass—all edges of the mold must be filed or sanded to take off the razor sharpness.

Plastilene—Plastilene is used for either small or simple one-time-use molds of objects having no undercuts. It's fine for simple surface impressions. Simply press it evenly onto the object,

making sure that you get a good impression, then peel carefully—you should have a perfect negative of the object, ready for casting. Plastilene gives fair detail, doesn't shrink and is reusable and resonably priced. However, it has poor release ability and may adhere to the object you're molding. It tends to tear and crack if not peeled carefully. Don't handle it too much, because the heat of your hands will soften it and it will be even harder to remove. A little trick in making simple sheet molds is to soften the plastilene in your hands, press it into place and then cool it while it is still on the object. If it is small enough, put it in the refrigerator for ten minutes or so, then peel. If it is not possible to put the mold in the refrigerator, let it stand for an hour or until it feels harder to the touch. Once it firms up, you will be able to remove it from the original with less tearing and cracking.

Plastilene is available in most art-supply stores and comes in hard, medium and soft grades. Medium is the best all-around grade for restoration molding. The soft grade might catch in the detail and rip, or might not hold detail; the hard grade may crack off when you try to remove it.

Casting Materials

Plaster of Paris—Plaster is a fine casting material, and most of the advantages of molding with plaster pertain to casting as well. Plaster can be cast in most molding compounds—even in plaster molds—if you use a mold-release agent. The amount of plaster cast is critical, however. Too heavy a piece may require additional mold support and too thin a piece may be fragile.

If you cast plaster in a plaster mold, here are a few tips. Always use a mold release. Coat all surfaces of the mold, the inside as well as the outside, where plaster might drip and stick. I recommend purchasing inexpensive, commercially

prepared mold release, but you can make your own mold release for plaster by boiling a pound of Ivory soap in about two quarts of water until dissolved, then adding two more quarts of water to thin. Let the brew settle overnight. Thick, glutinous sizing will settle to the bottom, while a clear, syrupy liquid will be at the top. This sizing will be just right to use. Remember, if you don't size the mold before you pour a cast, all you will have is a large plaster doorstop.

A variety of products having characteristics similar to plaster are available for use in casting. Hydrostone, for example, is a mixture of very fine plaster and ground stone. It is stronger and gives finer detail than plaster, and has a sparkly, stonelike look. It comes in terra cotta, black, brown, French buff, grey-green, bronze and crystal white, and you can mix your own colors by adding water-based pigments. Hydrocal, less expensive than Hydrostone (both are more expensive than plaster), gives a dead-white, matte surface. Hydrocal is also stronger than plaster, and gives finer detail.

Latex casting rubber—Despite being sold under a different name, this material seems identical to liquid rubber latex. It is also used with filler, but usually more than is used for moldmaking: Three parts filler to one part latex casting rubber gives a good, rigid cast. Latex casts are flexible and can be cut or trimmed with a sharp knife or razor blade. They can be given a patina or antiqued, and are light enough to mount with white glue (such as Elmer's) or, for a more permanent bond, epoxy adhesive. Latex casting rubber is inexpensive and easy to use in most molds with the proper release agent.

Monzini liquid casting compounds—Monzini compounds have been used by sculptors for

---

Just wiggle your fingers

How you stir plaster for casting is important, because you don't want air bubbles trapped in the mixture. This is how I learned to do it. Sprinkle plaster in the center of a bowl filled with water (never the other way around). Keep sprinkling until a mound of plaster just rises out of the water. Then put your hand down to the bottom of the bowl and wiggle your fingers so that the plaster is stirred from the bottom and no air bubbles are entrapped. When the plaster has a smooth, slightly-more-than-heavy-cream consistency, you are ready to pour. Plaster is curious stuff: If you sift it into water and don't disturb it, it will not set up quickly. Once mixed it begins to set up normally.

Pour plaster smoothly, without splashing, to avoid trapping air bubbles in the cast. After pouring, but before the plaster sets, vibrate the mold by pounding the table beneath with a rubber mallet, or by tapping the mold sharply a few times. —*J.T.*

years to simulate the textures of bronze, Carrara marble and aluminum. Monzini Bonded Bronze consists of bits of pulverized bronze fused together with bonding liquid. You can cast it in a mold, or apply it to the surface of a previously cast object for a bronze appearance. Carrazini Monzini, the most popular of the compounds, is ivory-colored. To produce the grains and colors of marble, Monzini color pastes are added. Lumizini produces casts that both look like and can be polished like aluminum. All the compounds will reproduce detail as fine as an object's finish, whether matte or polished. Monzini compounds set in under 30 minutes at 75°F, and can be cast in plaster, latex rubber, RTV, metal, wood or Monzini molds, though because of the expense,

casting Monzini in Monzini is impractical. Defects or air bubbles can be repaired simply by filling them with the same material. Monzini compounds won't shrink, and the casts are weatherproof. They can be machined, sanded, drilled, taped and polished. You can weld pieces together using more Monzini compound and clamping the pieces until they set.

Monzini compounds are expensive but worth it. Use Monzini compounds within three months from the date on the label, and be sure to use up however much you mix.

Low-melt metals—The Kindt-Collins Co. distributes Cerro Alloys, a line of six metals with low melting points.

Melt low-melt metals in a double boiler on the kitchen stove. Be sure to use an old pot that you don't need for cooking food, because alloy residue may be difficult to remove completely. Pour the molten metal into your mold and let it cool, and you have a metal cast without going to a foundry. But this convenience doesn't come cheap: Cerro Alloys are sold in two-pound lots, priced by the pound. Prices range from about $10 per pound for Cerrobase 255°F to about $19 for Cerrotru 218°F. A hint for casting low-melt metals: Powder the mold with graphite so the casting pops out of the mold easily.

Polyurethane foam—This material is appropriate where lightweight, bulky castings are

A

B

C

Molding and casting a ceiling bracket

Clean and repair the object to be reproduced and seal its edges with clay **(A)**. Then shellac it, and let it dry. Paint on a coat of fillerless mold rubber, leaving a lip at least 1 in. wide around the object. Brush out any air bubbles. When the rubber has dried, apply the second coat. Let it dry. Mix mold rubber and filler 1:1 in a clean container. Paint on two coats of this mix, drying in between.

Next lay on strips of cheesecloth 2 in. wide by 3 in. to 5 in. long, covering with rubber/filler mix as you go. Cover the object with two or three layers of cheesecloth and rubber/filler **(B)**, allowing drying time between, and finish with a coat of rubber/filler. Let dry, then trim the lip with a mat knife.

Because the weight of a casting can pull even a reinforced mold out of shape, you'll need a mother mold (usually made out of plaster) to support the rubber mold during casting. A simple casting jig made of four pieces of wood and angle irons (see drawing on facing page) will hold the mother mold.

If the object has undercuts, you'll need a two-part mother mold. First, make a shim by cutting a piece of cardboard to the shape of the mold. In places the cardboard won't follow the form, tape pieces of scrap in position and trim to fit. Place this guide on a new piece of heavy cardboard and trace the shape to make the shim. To prevent plaster from running through to the other side, fill open spaces in the shim with clay **(C)**. Apply clay to the side of the shim opposite the side you'll be pouring first.

Place mold and shim in the casting jig, shoring up the sides of the shim with wood supports clamped to the jig with C-clamps. Tape all edges with masking tape to prevent the plaster from leaking into the other side **(D)**. Size the inside of the casting box, the mold and the shim with mold release.

Mix the needed amount of plaster to a heavy-cream consistency. Pour the plaster into the jig until it's an inch or two above the mold **(E)**. Let it set, then remove the shim.

Scoop out three alignment holes **(F)** with plaster modeling tools or a hooked knife. Then size the mold, casting box and mother mold so

D

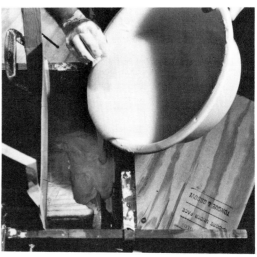

E

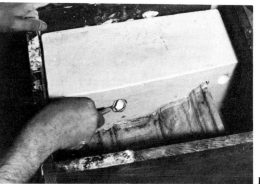

F

needed, such as fake wooden beams and simulated moldings. It can be painted, given a patina, and cut and shaped.

Polyurethane foam is a two-part casting medium that, when mixed, quickly expands approximately 30 times its liquid volume, filling the mold with foam. It's tack-free in about five minutes, and can be removed from the mold in about twenty. You can cast polyurethane foam in plaster, RTV, latex, wooden or metal molds treated with a mold release. Polyurethane foams are available in a range of colors and in three weights: light, medium and dense.

Epoxy casting resin—There are so many different manufacturers and types of epoxy casting resin that I will touch on only a few of specific interest to the restorer. Emerson and Cuming, Inc., for example, produces the Stycast line of about 70 casting resins, commonly used in the electronics industry. These include general-purpose casting resin (Stycast 2651); a foam epoxy used in aircraft that is so light it floats (Stycast 1090), and a transparent epoxy gel that, when cured, is tough but flexible (Eccogel 1265).

High detail casts, little shrinkage, low cost and ease of use are some of its advantages. Epoxy-resin castings can be machine-cut, drilled and taped; they also adhere to metals, plastics and ceramics. Disadvantages include short pot lives, poor acceptance of paint, and for some, vacuum evacuation for bubble-free casts. ☐

Sources of supply

Silastic RTV silicone rubber and series 31 RTV rubber: Dow Corning Corp., P.O. Box 1767, Midland, Mich. 48640, also Industrial Plastics Supply, 309 Canal St., New York, N.Y. 10013.

Hydrocal, Hydrostone, moulage, liquid rubber latex, latex casting rubber: Sculpture House, 38 E. 30th St., New York, N.Y. 10016 or Arthur Brown Bros., Inc., 2 W. 46th St., New York, N.Y. 10036.

Adrub RTV rubber, Arte-Vee RTV Kwikset, Adrub RTV Softee, Monzini compounds, polyurethane foam: Adhesive Products Corp., 1660 Boone Ave., Bronx, N.Y. 10460; Industrial Plastics Supply.

Eccosil RTV silicone rubbers, Stycast epoxy casting resins: Emerson & Cuming, Inc., Canton, Mass. 02021.

Low-melt metals: The Kindt-Collins Co., 12651 Elmwood Ave., Cleveland, Ohio 44111.

Cab-O-Sil: Industrial Plastics Supply.

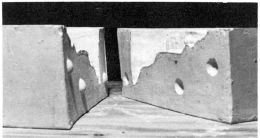

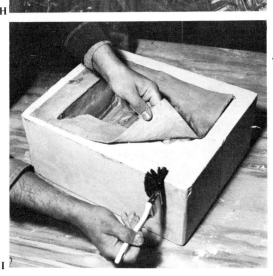

that the new plaster will not stick. Pour the other side and let set.

Pry mating halves of the mother mold apart with a spatula. Try to let both sides dry thoroughly before casting **(G)**.

Carefully peel back the latex mold from the original object **(H)**. Scrub the mold to get rid of any bits of debris that might have accumulated inside. (Liquid rubber latex is notoriously rough on surface patinas.) Dip the mold in very hot water to vulcanize and it's ready for use. (If you don't use the mold until sometime later and find it has become rigid, simply immerse it in hot water until it regains flexibility.)

Place the mold in the mother mold, sizing both mold and mother mold with mold release **(I)**. Mix enough plaster for one cast and pour, breaking all the bubbles. Work quickly, and fill the mold before the plaster sets **(J)**. Tap the table with a mallet to help release air bubbles from the cast, and smooth out the back of the cast with a straightedge **(K)**. Now is the time, while the plaster is setting, to add a hanger or bracket, if necessary for mounting.

When the cast is dry, simply lift it and the mold out of the mother mold and pull the rubber mold back **(L)**. Trim off any excess from around the foot of the cast.

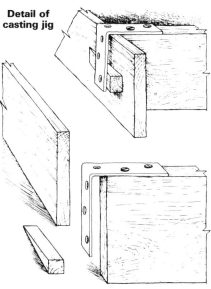

Detail of casting jig

All corners slip together and are held tight by wooden wedges.

Drywall Finishing

One contractor's techniques, using a taping banjo and stilts

by Craig Stead

It was a hot July in 1977, and I was working on a run-down two-story house. Much of the work involved hiding ugly cracked and crumbling plaster walls and ceilings with sheetrock. I was over budget and late on schedule. Large-scale drywall finishing was rumored to be an esoteric art practiced only by initiates, so I decided to subcontract it out. I started phoning finishers listed in the Yellow Pages. Fifteen calls later, I had not a single bid on the work. At that point, I was getting a sinking feeling in my stomach because I could see the whole project coming unglued. My client was eager to move in. The crew was standing around waiting to start the painting. Desperate action was needed. I popped down to the local lumberyard, bought everything that looked related to drywall finishing, grabbed a stack of sandpaper, and set to work. The going was slow, the sanding endless, the dust every-

where. I found that with enough time and sanding, you can achieve an acceptable finish. I also found that I hated sanding drywall seams. I got the job done, and vowed that I would never get caught in that bind again.

Since then, I have finished the drywall on all my jobs; and I have talked to every pro who would give me the time. I even worked one house alongside a drywall contractor to learn the tricks that are so important to speed and final quality. I found that professional rockers don't sand much. In fact, they don't sand at all until the last coat has gone on. The professional approach is to use the finishing knife or trowel to smooth the drywall mud while it's wet so the coat is free of ridges and high spots.

There are specialized tools for drywall work. In some parts of the country, sheetrock contractors use bazookas (they look a little

like anti-tank weapons). In the time it takes to walk the length of a joint, a bazooka can apply the tape, smooth it down and cut it off. Bazookas are typically used with "boxing" equipment that lays down properly feathered fill and finish coats without using a trowel. For years, bazookas could only be leased, and the charge was well over $1,000 a month. Even though they can now be purchased, they cost a lot and only high-volume drywall contractors would want them. If you're a professional who builds on a large scale or a novice with limited time and expertise, you may find it cheaper and quicker to call one of these well-equipped specialists. But if you're out in the boonies, or you are a small contractor with time between jobs, you're probably better off doing the work yourself.

I do a lot of renovation, which most drywall contractors don't like to touch. For me, alumi-

From *Fine Homebuilding* magazine (October 1984) 23:60-65

num stilts and a a taping banjo like the one I'm using in the photo, facing page—a metal box that holds mud and a 500 ft. of tape—were investments worth making as a general contractor. A banjo isn't as slick and quick as a bazooka, but it costs a lot less, and it helps me do a good job with some speed. I'll be talking more about the banjo and stilts below.

Hanging the rock—The first step to a quality drywall finishing job is hanging the sheets right. You've got to have flat walls, solid nailing and correctly installed outside corner beads. Most of this is discussed in Bob Syvanen's article "Drywall" (see pp. 90-95), but I'll briefly review some pertinent points here.

Drywall comes with its long edges tapered, and its short edges the full thickness of the rock. The sheets are hung so that the full-thickness edges butt on studs. The tapered edges span between studs. Butt seams should be staggered from one course to the next so they don't all line up. If you have a tapered edge meeting a butt edge, you have a mismatch that is difficult to finish.

For drywall, walls should be framed with surfaced, kiln-dried lumber. In remodeling, where you often uncover walls built of rough cut lumber, you have to shim the studs to get them all to align. I always check the walls and ceilings on my remodel jobs with a 6-ft. aluminum level or a string line. I correct small variations with the $\frac{1}{16}$-in. cardboard drywall shims that are commonly available around Boulder, Colo., where I used to work, or with 4-ft. long shims that I rip on my table saw in $\frac{1}{16}$-in. increments from $\frac{1}{16}$ in. to $\frac{3}{4}$ in. thick. For a ceiling that's in bad shape, I either strap the whole thing with 1x4s, leveling them with shim shingles, or I frame a drop ceiling using 2-in. lumber. This is a tedious job, but if you don't take the time to do it, your final results will suffer.

Once you have flush framing, the hanging hints in the box at right can save you time.

Materials—To get a high-quality finishing job, you will need paper joint tape and premixed drywall finishing compound. I also use a filler called Durabond 90, made by U. S. Gypsum Corp. (101 S. Wacker Dr., Chicago, Ill. 60606). The paper tape bridges the joints between the sheets. Fiberglass mesh tape with an adhesive backing is also available, but it's much more expensive than the paper tape and is more typically used with veneer plasterboard (blueboard).

Drywall mud, or joint compound, is both an adhesive and a filling material. You use it to glue the tape over the seam, and then you fill the seam with it. It sets up as the water in it evaporates. After it dries, it can be redissolved to a certain extent by water. This means that you can smooth and feather a joint by washing it with a sponge if you need to. Tool cleanup is simple, because dried compound soaks off with water.

Don't dilute premixed drywall mud unless you're using it in the taping banjo. I mix each new pail of mud with a power paint mixer in a

$\frac{1}{2}$-in. drill, just enough to give the mud a smooth, even texture. I keep the pail covered at all times so the mud won't dry out. As I use the mud, I continually clean down the sides of the pail with a drywall knife to keep the thin side layers of mud from drying out and falling into the bucket. These lumps create dragouts (trenches) in the fill coat when you are smoothing the seam. If you get a dragout, clean your knife on your mudpan and continue smoothing the seam.

I keep a 6-in. drywall knife stuck in my mud pail to clean the bucket sides and to fill the mudpans and banjo. Each evening I empty my mudpans into the pail, then smooth the top of the mud. If I'm storing a partially filled bucket of mud for a long period of time, I lay a sheet of plastic over the surface or pour $\frac{1}{4}$ in. of water over it to keep the mud from drying out. Keeping your mud pail and tools clean goes a

long way toward creating a satisfactory finishing job.

Durabond 90 is similar to patching plaster. It sets up by chemical reaction (hydration) rather than by evaporation. Once it's dry, it is not water soluble, does not shrink, and is hard to sand. It comes as a powder, and you mix only as much of it on the job site as you can use before it hardens (it has a 90-minute working time). You must also clean your tools immediately, or the stuff will stick hard and be difficult to get off.

I learned about Durabond 90 in Colorado from a fellow on my crew who had spent two years doing nothing but drywall. His outfit used it a lot to fill cracks, and sometimes as the fill coat, although most drywallers use only joint compound. I think the stuff is well worth the trouble of mixing it. I use it to fill all large gaps in joints after the rock has been hung. It's good for filling gapped corners, holes, and the crevice formed where a kneewall intersects a cathedral ceiling. On my jobs, any gap greater than $\frac{1}{4}$ in. gets a Durabond fill. Otherwise, the mud can suck the tape into the crack. This is particularly bothersome on inside corners. When I'm trying to move quickly on a job, I often use Durabond 90 for the first fill coat on outside corners and tapered joints because I can go back and get a coat of mud over it in an hour and a half instead of having to wait until the next day. The important thing to remember when you use the material this way is not to overfill a seam. It's hard to sand down high spots.

Gypsolite compound (Gold Bond Div. of National Gypsum, Charlotte, N.C.) is useful for heavy fills, like patching a large hole in a plaster wall. This is brown-coat gypsum plaster with perlite added. It resists cracking in thick sections, and it's inexpensive. But you have to overcoat it because it is grainy. It is not often used for drywall finishing, but it can be useful in remodeling work when the rock is running to an old plaster wall. In these situations, you can have voids as deep as several inches, and Gypsolite is just the ticket for filling them.

The taping banjo—For the kind of work I do, the taping banjo (photo facing page) is a terrific tool. It puts a layer of thinned drywall mud on the tape as you lay the tape on the joint, so you just smooth it down with your knife, saving the step of mudding the joint before applying the tape. The nose of the banjo has a toothed edge for cutting the tape. It's a great time-saver, and with it I can tape off a three-bedroom house in a little over a day, about a third the time it takes with taping knives alone (with a bazooka, the same job might take four to six hours).

Two types of banjos are available: dry and wet. In the dry type, the roll of tape is kept separate from the mudbox, which holds the thinned mud. In the wet type, the roll of tape is mounted inside the mudbox. Mine is a dry type made by Marshalltown (Marshalltown Trowel Co., Box 738, Marshalltown, Iowa 50158). It works well. I have never tried the wet type. Banjos are made of aluminum or

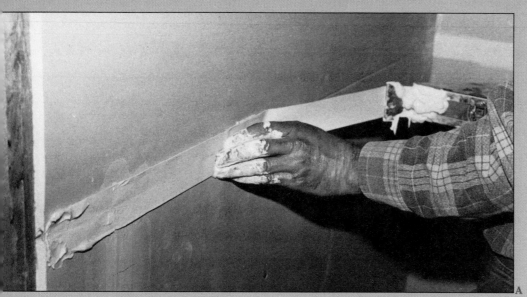

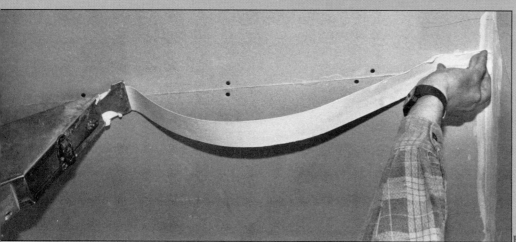

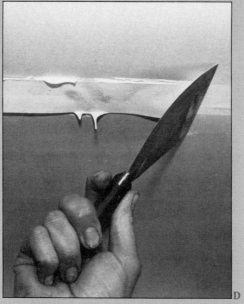

Taping seams. **A taping banjo (A) lays the mudded tape along the joints between sheets of drywall. You pull tape out of the tool's throat to get it started, then press the tape to the wall every foot or so with your hand to hold it in place. On ceilings, tape is liable to fall off, so once Stead gets it started, he lets a bit sag (B). Then he tugs the banjo sharply and snaps the tape up against the drywall. The force of the snap holds the tape in place until it's smoothed down with a knife. At corners, it's important to keep the fold centered, and press the tape snug every 5 in. or so (C). Once the tape is applied, it's smoothed into place with a 6-in. knife (D).**

stainless steel. I have used both types and prefer the aluminum model because it weighs a lot less.

To set up a dry banjo, you first install the roll of tape in the holder. I use 500-ft. rolls, which save reloading time. A typical three-bedroom house requires five to six rolls of tape. Next, you thin your mud to the consistency of condensed tomato soup—about ½ gal. of water per 5-gal. pail. If the mud is too thin, it will drip out of the nose of the banjo; if the mud is too thick, the joint tape will be hard to pull through, and may break.

Next, set the adjustable slide in the nose of the banjo so that it deposits a ¹⁄₁₆-in. thick layer of mud on the tape. This minimizes squeeze-out when you smooth the tape down with your knife. Now loop the tape through the mudbox of the banjo, as shown in the photo, facing page, left.

Load the mudbox with your thinned mud and you are ready to go. Before you begin, grease your hands with Vaseline. You'll be getting mud all over them, and this keeps them from drying out.

The order in which you tape the seams is important. Do all the butt seams first, the tapered seams second, and the inside corners last. Lap your tape. The butt-seam tapes should go from, say, a corner to the center of a tapered seam. The tape on the tapered seams and corner tape will then lap over the ends of the tape on the butt seams. This way, no tape ends will be torn loose by the finish knife, and the corners will come up smoothly when troweled.

Pull the tape out of the banjo and press it to the seam with your free hand (photo A, top left). You can press it every foot or so just to hold the tape to the seam until you smooth it down with your knife. Once you have run the tape over the seam, set the banjo down and smooth the tape down with a 6-in. knife (D). You want that tape tight, particularly on the butt seams. Clean the excess mud on your knife into the mudpan and return this mud to the banjo for reuse. Wipe both sides of the joint clean of ridges so you have a well-bedded tape with no streaks of mud on either side. On tapered seams, the tape will not lie as tight because you are smoothing it into the dip between the two sheets. Don't worry about this—a heavy fill coat will cover up the tape later.

Slightly mismatched butt seams where one side is higher than the other can produce tape bubbles. To avoid these, smooth the low half of the tape first, and then smooth the high side, creating a step effect with the tape. Don't worry if it looks funny; a crown trowel—a 12-in. trowel that is slightly curved along its length—will take care of the problem when the fill coat is applied.

If you forget to smooth down a tape and it dries to a wrinkled mess on the seam, wet the tape with a sponge periodically to soften the mud. Then scrape off the tape with a drywall knife and try again.

If I'm working alone, I tape four or five seams with the banjo before doing the

Taping banjo and stilts. **At the core of the author's drywall technique, the dry banjo (above) holds a 500-ft. roll of tape and dispenses it through a chamber filled with mud. The stilts (right) are considered so dangerous by some agencies that they are banned in some states. Stead feels that they can be effective for doing high work if they're used with care.**

smooth-down. This reduces the number of times I have to switch from the banjo to the smoothing knife and mudpan. For large jobs, the most efficient procedure is to have one person running the banjo and another smoothing the tape. On ceilings, pull out about 3 ft. of tape, allow it to sag 3 in. or so from the ceiling, and then quickly pull the banjo to snap the loop of tape to the ceiling (B, facing page). This tug slaps the mud side of the tape to the ceiling and holds it for its full length. Otherwise, the tape may peel off and drop in a sticky mess on the floor.

Inside corners are done last and require some fussing to get good results. I tape the vertical corners first, from ceiling to the floor, and then do the horizontal ceiling corners. What you want is a smooth, continuous tape that exactly meets the corner at either end. If the tape is too short, it leaves a hole in the corner. If it's too long, it piles up in the corner and must be trimmed to fit. If you do cut the tape short, splice in another piece. Any overrun can be cut back with a utility knife.

As you are pulling the tape from the banjo, tuck it into the corner centered on the fold, and press it in about every 5 in. (C, facing page). Smooth the corner tape carefully with your 6-in. knife, and try to form a wrinkle-free 90° angle. Your corner trowel is going to ride on this tape during the fill coat—some care here will make quality easier to achieve with your final coat.

The banjo is difficult to clean. If you are going to be using it for several days, store it nose down in a pail of water at night. This keeps the tape from drying out and sticking to the nose of the banjo. When the job is over, hose out the banjo and use the scrub brush to get into the corners of the mudbox.

Tools for finishing—I use a mudpan and drywall knife once the tape is on the wall. I also use an inside-corner trowel, and a crown trowel for finishing butt seams and problem seams. I prefer a mudpan to a hawk because it's easier to clean knives on the edge of it, and because it exposes less of the mud to the air so the mud doesn't dry out so quickly.

I have finishing knives from 6 in. wide to 12 in. wide in 2-in. increments. The 12-incher is the largest knife that will fit my mudpan. I like Harrington knives (Harrington Tools, Box 39879, Los Angeles, Calif. 90039) because they are more flexible than other brands, and I can feather the final mud coat better with them than with a stiffer knife. Every once in a while, I want to feather a seam real wide, or coat two neighboring seams at once. For this special occasion, I have an 18-in. wide knife.

I carry a hammer on my belt when laying the first coat of mud to set high nails. Tool cleanup is always needed on a drywall finishing job, so I put a mud pail filled with water in the center of the floor. In the pail floats a stiff tire scrub brush for brushing off hardened mud. All my equipment, the water pail plus the mud I'm using, sits on a sheet of plywood or tin to protect the floor from spills.

Stilts—In some states, safety regulations don't allow the use of stilts and will not award Workman's Compensation to anyone injured using them on the job. Many people, including a lot of professional tapers, feel that stilts are among the most dangerous tools around. A fall can pop your knee or snap your leg. I feel, however, that stilts are an effective way to handle high work if you're reasonably careful. I use stilts to finish ceilings and upper walls. With them, you can get smooth ceiling seams

without sanding because you can smooth the seams with one continuous motion of your finishing knife. They do require some practice to use comfortably and safely.

Before you start your drywall finishing, clean up the area thoroughly. Get rid of all debris, scrap lumber, and leftover rock. You're liable to trip over these things when you are doing overhead finishing on stilts. Besides cleaning up the area, I stack up drywall pails so I can refill my mudpan conveniently when I'm up in the air.

Various brands of stilts are available; the ones I use are made by Dura-Stilt Corp. (8316 S.W. 8th, Oklahoma City, Okla. 73128) and cost around $180. When you order a pair of stilts, get the accessory that lets you strap them to your boots. Otherwise you have to mount your boots permanently to the stilts with screws. Make sure your stilts are in good condition and inspect them before every use. Adjust the stilts to your weight and body size as recommended in the manufacturer's literature. Stilts that are out of adjustment can tire you out, and make you more liable to fall. If you are new to stilts, or haven't worked with them for a while, take some time to get the feel of them by walking around the room near the walls so if you start to fall you can catch yourself on the wall. Mount your stilts leaning against a wall, and attach the upper leg strap first before you fasten the boot straps (photo, above right). Going up and down stairs on stilts is not recommended, though I know people who do it. I scaffold stair halls and don't use my stilts on the stairs.

The fill coat—Finishing drywall is like a striptease—you put it on and then you take it off. You apply a coat of mud, smooth it out

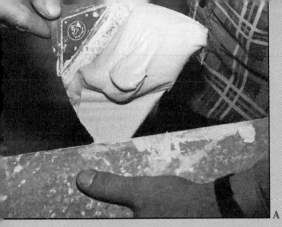

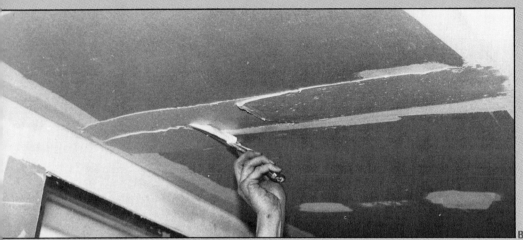

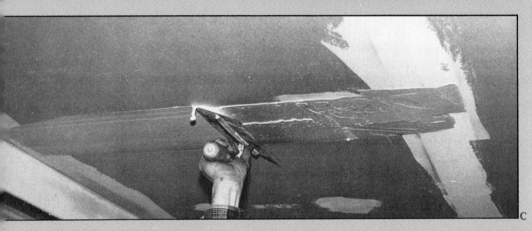

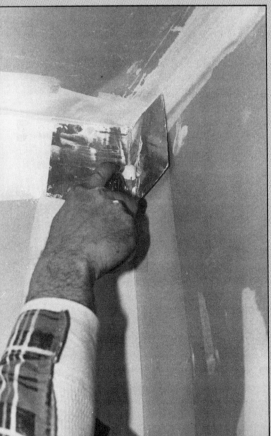

Applying the fill coat. To cut down on squeeze-out when he's applying mud to a seam, the author uses the sides of his mudpan to cut the compound back on the corner of his knife (A). Stead tries to avoid butt seams, but it's virtually impossible to do a drywall job without having to deal with a few. To fill butt seams, a separate band of mud is laid on each side of the joint (B), then smoothed down with a crown trowel (C). Filling inside corners is tricky, too. Stead lays a band of mud on each side of the corner, then smooths them down with a corner trowel (D) before smoothing the remaining mud ridges on each side of the corner out with a 6-in. knife (E).

with the finishing knife by removing some of the layer, and clean your knife on the edge of the mudpan with each stroke. Using the knife to smooth and feather the wet mud is easier and faster than sanding the stuff when it dries. This approach avoids a lot of dust, and you don't have problems with fuzzing the paper covering of the drywall. Any small ridges or blips of dry mud that need removing between mud coats are knocked off with the edge of a finishing knife.

I apply two or three coats of mud after I lay the joint tape on with the banjo. A systematic approach to the application of each coat of mud is the key. You must do things in order and not bounce around the room, wasting time and energy. The principles are the same in covering all those nails or screws—three coats of mud, generously applied and smoothed off level with the surface of the sheetrock (assuming you want smooth, not textured, walls).

The first finishing coat of mud is called the fill coat. It is thick, so to make sure it dries in a reasonable time, keep the temperature of the room between 70°F and 75°F. As the fill coat dries, it shrinks, and looks pretty bad. Novices are tempted to start sanding immediately. Don't.

When you apply mud with a knife to a seam, the mud tends to ooze off its ends as you are laying it on and then drips onto the floor. Cutting back the mud on the corners of the knife before applying it substantially reduces this problem (A).

Fill the tapered seams first, the butts second, the outside corners third, and the inside corners last. The tapered seams are filled first because you want to get the surface flat before dealing with the other types of seams. Lay on your mud with a 6-in. knife and smooth the joint with an 8-in. knife. Be sure you completely fill the depression to the surface. In smoothing the joint, keep the knife almost perpendicular to the wall, and use light pressure on the knife. Too sharp a knife angle coupled with a lot of pressure can curve the blade and dish the joint.

In Colorado, I often used drywall shims under the rock one stud back in each direction from butt joints. This created a 1/16-in. dip at the joints, which I treated as if they were factory-tapered. Since the cardboard shims are not available in Vermont, I lay two bands of mud on butt seams, one to either side of the tape, and smooth the seam with a crown trowel to cover the tape with mud and to establish a knife guide for the next coat (B and C). Here I want a good layer of mud over the tape. Don't worry if the edges of mud farthest from the seam are a little ragged. The next coat will take care of them.

Outside corners take a lot of mud on the fill coat. If I'm filling a lot of them, I lay on the mud with a 12-in. knife and apply the mud with the handle of the knife held parallel to the floor. Then I smooth down each side of the corner using an 8-in. knife. Make sure you smooth the mud that goes around the corner and gets on the other side. If you forget to do

Finishing. To finish inside corners, Stead lays a thin band of mud with a 6-in. knife over the remaining ridges (above), then smooths them down with an 8-in. knife (right).

this, you can knock the dry globs off with your knife before you do the next coat.

Inside corners are the trickiest to fill. Lay a band of mud to each side of the corner using a 6-in. knife, making sure that you get the mud fully into the corner. Then smooth the corner down using a corner trowel (D, facing page). You may have to smooth it several times to get it good and clean. Make sure you start the trowel tight into your starting point so the intersections of the two walls and the ceiling come up clean. After running your corner trowel, smooth the mud ridges left on the wall with your 6-in. knife (E, facing page).

If you pull your knife off the seam as you are smoothing the joint, it will leave a ridge of mud that may need sanding when it dries. Try to trowel your joints in one smooth, continuous motion. If you have to break the knife off of the joint, do so where another joint crosses the seam. If you are smoothing a tapered seam, lift your knife as you come to a butt seam that intersects it. This way your next coat on the butt seam will cover the ridge on the tapered seam.

The finish coats—The finishing coats are thin. They dry in three or four hours, and shrink very little. The final coat is almost ironed on and is thinner than a dime.

The order of seams is not important on this coat; just make sure that you do all of one type of seam before jumping to the next type. That way, you are unlikely to cross a wet seam as you proceed around the room. I use the same order of seams as in the fill coat, or put one worker on one type of seam while another does the corners.

For tapered seams, I lay on the mud with an 8-in. knife and smooth it down with a 10-in. knife. Butt seams are done the same way, only two bands of mud are laid down, on either side of the tape, and then smoothed down in two passes. On outside corners, I use a 12-in. knife to lay on the mud as before and then smooth it down with a 10-in. knife. Inside corners are fun at this point because they merely need a thin band of mud laid on with a 6-in.

knife where the ridge was left by the corner trowel. Then this is smoothed down with an 8-in. knife (photos above).

Smoothing a seam requires that your knife have two smooth guides to run on. For example, an outside corner is covered with metal bead not only to give it strength and protection from impact, but also to provide a guide for the finishing knife. In smoothing such a corner, your knife rides on the corner bead on one side, and the smooth surface of the drywall on the other. On any seam I'm finishing, I'm looking for my smooth guides. On tapered seams, the smooth surfaces of the drywall on each side of the seam are the guides. On inside corners, the corner tape and the drywall surface are the guides. That crisp, straight corner tape is one of your knife guides for feathering the coats of mud.

Butt seams are the greatest problem because the center where you lay your tape is a high point and the guides for the knife thus become the center of the tape and the smooth drywall to either side of the seam. Butt-seam tapes are often wrinkled from the nails underneath them, and the smoothing knife chatters on the lumpy tape. One approach to finishing a butt seam is to lay two bands of mud on it, with the tape in the center of the two bands, leaving a ridge to sand later. This works, but you must have a lot of knife control as you are working with only one knife guide and floating the center of the seam. Commonly, you get either too thick a coat, which shows as a bulge over the seam, or you starve the seam for mud, get a very thin mud coat over the tape, and break through the mud to the tape when you sand out the ridge. This leaves a fuzzy tape, which shows after you paint. My method is to use the crowned bed of mud as the guide for the knife, and subsequent coats of mud completely hide the seam.

The second and final finish coat is very thin. It conceals minor imperfections. It is smoothed off with a lot of pressure on the knife, and if it's done right, it feathers out to almost no ridge on the seam. I lay on and smooth off the mud by running the knife with

the seam at a 45° angle, or crosswise. On the final coat, I lay on my mud with a 10-in. knife, and smooth with my 12-in. knife. Keep smoothing the joint until you have no ridges of mud left, just an even, well-ironed, thin coat of mud. This coat dries in about an hour and goes on quite quickly. The order of the seams is not important. Some seams give you a chatter or washboard effect. If you encounter this, lay your mud on crosswise to the chatter, and smooth it off in the same direction.

Some light sanding and sponging of the seams are all you need to complete the finishing job. Inspect the seams as you work, and circle any areas needing touchup with a pencil so you can correct them before painting. A strong light on an extension cord is useful for finding problem areas in hallways and dark corners. Also inspect your work after the first coat of paint is applied; sometimes previously unseen imperfections become noticeable when the surface is all one color.

If I haven't done any drywall finishing for a while or if I'm training someone new, I start in the closets to warm up on techniques. Work your back rooms first and save the living room and dining room for last. That way your best work will be where people notice it the most. The new worker gets the nails as a starter to get the feel of the 6-in. knife, the drywall mud and the mudpan. My rule is that you always sand your own work so you can get feedback on how you are smoothing the joints. A common problem with novice workers is oversanding tapered seams, leaving them slightly depressed.

There isn't one way to finish a drywall job, just a final result that must look right. Your corners must be clean and crisp, especially where the ceiling meets the intersection of the two walls; the tapered seams must be fully filled and flush with the surface of the rock, and the butt seams must be feathered out far enough to each side so they don't show as a bulge under raking rays of light. That is your final test and grade on your finishing work. □

Craig Stead is a builder in Putney, Vt.

Drywall

Hanging and finishing gypboard can be an aggravating mess. A veteran builder shows how to do it right the first time

by Bob Syvanen

Gypsum board, gypboard, Sheetrock (a brand name), drywall—whatever it's called, it's probably the most disliked and misunderstood stuff in house building. The sheets are heavy, the sanding seems endless and the dust is unpleasant. But a drywall job well done is very satisfying. There are many ways to do it, but I have settled on this system after years of trial and error, watching others and asking questions.

Drywall comes in sheets from 4x6 to 4x16 and in ¼-in., ⅜-in., ½-in., and ⅝-in. thicknesses. Quarter-inch drywall is used over old plaster walls so it is always fully backed; ⅜-in. is for cheap construction, and the framing must be no greater than 16 in. o.c. You can use ⅜-in. material double, so you end up with a ¾-in. thick wall. The most commonly used drywall is ½ in. thick; ⅝-in. is for extra good work. There is also a ⅝-in. fire-rated sheet, usually required on a garage wall shared with the house. Use water-resistant sheets in bathrooms, especially around the tub and lavatory.

Most lumberyards carry drywall sheets, but I prefer to buy from a drywall supplier who has a truck with a hydraulic arm. This is the best thing to come down the pike since drywall itself. Loaded with sheets of gypboard, the arm will reach right in the door or window, and sometimes even to the second floor. You might have to take out a window sash, but it's a small price to pay for such a glorious convenience. For a small load delivery, you may have to wait until the truck has a full load and is heading your way. It's advisable to put the sheets into the rooms where they'll be hung and to keep the sheets to be used first on top of the pile.

Editor's note: Bob Syvanen, a draftsman and carpenter in Brewster, Mass., is consulting editor to Fine Homebuilding. *This article is an expanded version of a section in his book* Interior Finish *($7.95 from East Woods Press, 820 E. Boulevard, Charlotte, N.C. 28203).*

Preparatory work

Before you begin hanging Sheetrock, it's best to clear the floors of tools and clutter. The sheets are just too heavy and cumbersome to carry through an obstacle course. A clean sweep-up is nice too, even though you'll shortly be up to your ears in debris again.

New England is the only place I know of where 1x3 strapping is used as a base for drywall ceilings. Although it's a lot of work, it's an excellent way to keep things flat with today's varying lumber sizes. Eyeball the ceiling for any bad joists (1) and trim with a Skilsaw before nailing up the strapping. You can also work to a stretched string with spacers at each end. Make minor adjustments with shims of wood shingle tips between the strapping and the joists (2). I like to double-nail for more holding power, but single-nailing works too. When a ceiling needs joints in the strapping, be sure to stagger them from one course to the next. Check the walls for bowed studs by holding a long, straight 2x4 against the wall (3). A badly bowed stud in a finished wall really shows, so replace or straighten it. To straighten a bowed stud, saw well into it on the side opposite the hump (4). The stud can then be forced into alignment. Wood shingle tips driven into the kerf will keep the stud straight while cleats are nailed on each side, just like splinting a broken bone together. It might take two such cuts if the bow is bad.

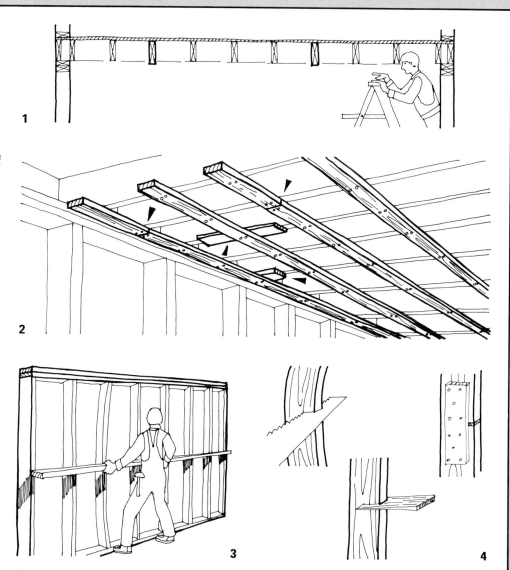

1

2

3

4

Ceiling joints

The fewer joints, the better. Ceilings are best done with a minimum of joints where the untapered edges of the sheets come together **(5)**. So where possible, span the width with a single sheet. Untapered edges make for bad joints **(6)**, but there are several ways to overcome this. Where the ceiling requires an untapered edge joint, it should fall between the joists or straps. The first solution is to nail up blocking 16 in. o.c. across a joist space and then a 2x4 or 1x3 strap parallel to the joists **(7)**, so that when the sheets are nailed they will be depressed about ⅛ in. Another way is to cut four 12-in. by 12-in. squares of Sheetrock. Butter these squares with joint compound and slip them in on the backside of the panel already in place **(8)**. Nail the next sheet up, place a piece of strapping along the seam and hold it in place with crosspieces of strapping **(9)**. These crosspieces will depress the joint, and the buttered squares will dry, holding everything in place like glue. Then remove the strapping braces, and tape and spackle the joint like any other. Although these methods are not commonly used by professionals, either one will result in a smoother surface, and require less work when these joints are taped and spackled.

I like to make a sketch of all the walls and ceilings to decide on sheet size and placement **(10)**. Sometimes I make my sketch on a stud or on a header or each wall, so I don't lose it. If you do the ceiling and butt the ends together, figure on using full sheets, plus the pieces cut to fit **(11)**. When working around the doors and windows, put the sheets in place first, and then cut out the openings. This is extra work, but if pieces are patched in over the doors and windows, they will crack at the seams as the framing shrinks and the house settles. The bumps at these joints won't let the casing lie flat at the corners without a lot of extra work **(12)**. If economy is your goal, figure the wall sheets any way you want, but if quality is what you want, run the sheets over the openings and then cut out.

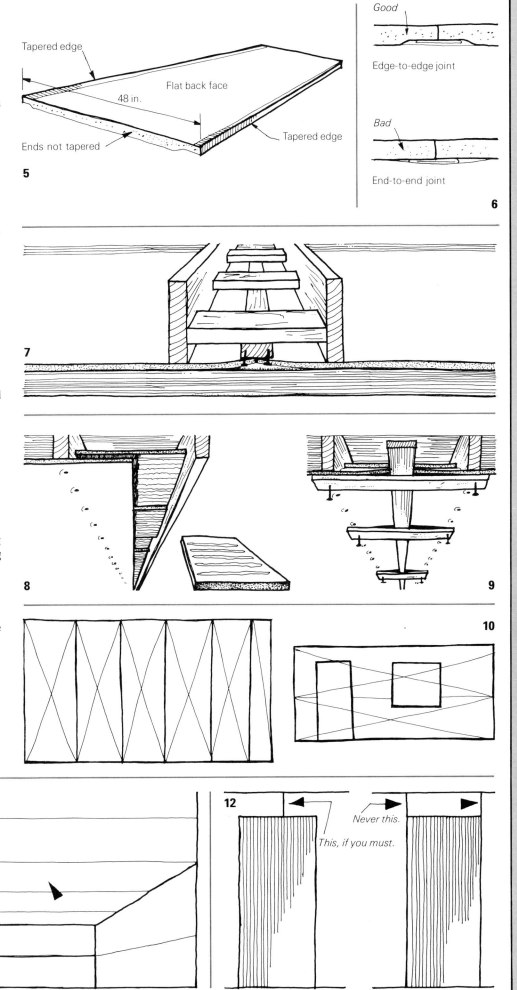

5

Tapered edge

Flat back face

48 in.

Ends not tapered

Tapered edge

6

Good

Edge-to-edge joint

Bad

End-to-end joint

7

8

9

10

11

12

Never this.

This, if you must.

Tools

For cutting drywall, a utility knife is useful. A 48-in. aluminum T-square is a must for laying out and guiding the cuts. A 12-in. keyhole saw **(13)** works best for cutting holes in the sheets. A Stanley Surform **(14)** is good for shaving a wallboard edge. A claw hammer works fine, but a drywall hammer **(15)** is lighter, and has a rounded head with a waffle pattern that leaves a dimple without tearing the paper. Use the blade to trim framing or as a jacking wedge. An aluminum hawk can be bought for about $7. A homemade plywood hawk works almost as well **(16)**. A galvanized metal or

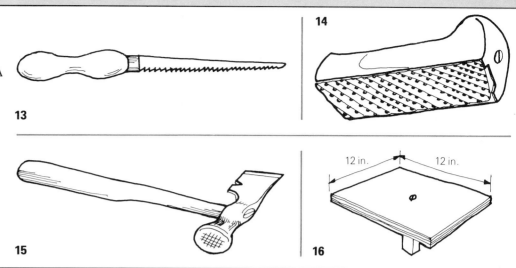

Fitting and nailing

The ceiling is the place to start, and you need two people to do it. I have done it alone, but don't recommend it at all. A ½-in. 4x12 sheet weighs almost 100 lb. A couple of T-braces will make the job easier. Be sure you can reach both of them and still control the sheet overhead. It also helps to stick a few nails in the panel where you'll be nailing. Holding the sheet and fishing for nails can be tough. The T-brace should be a little longer than the floor-to-ceiling height **(20)**, so that there is a slight wedging action. If it is too long, the brace won't stay in place, and if it's too short, the whole business comes down on top of you. The easiest way to hold a sheet against the ceiling is with your head. A sponge inside an old cap can make this a little less painful.

When nailing, be sure the sheet is pushed hard against the joist or strapping before driving the nail home. If it isn't, the nail will pull through the surface of the drywall sheet **(21)**. Once the paper surface is broken, the nail will not hold the sheet in place. If a nail misses a joist or stud, pull it out. Hit the hole with a hammer hard enough to depress the surface without breaking the paper. These dents, or dimples, can easily be filled with joint compound **(22)**. Just spot-nail the sheets enough to hold them in place at first. Secure them later. Cement-coated and annular-ring nails are okay for walls, but screws do a better fastening job on ceilings. Tool-rental shops have both the screws you'll need, and the driver. It's the size of a ¼-in. electric drill with a magnetized Phillips bit and an automatic depth stop. If you must use nails, a good way to get ceiling sheets to pull up is by double nailing **(23)**. First, single-nail 12 in. o.c., starting from the middle to prevent bulges from forming. Then come back with a second set of nails about 2 in. from the first set. You might have to drive

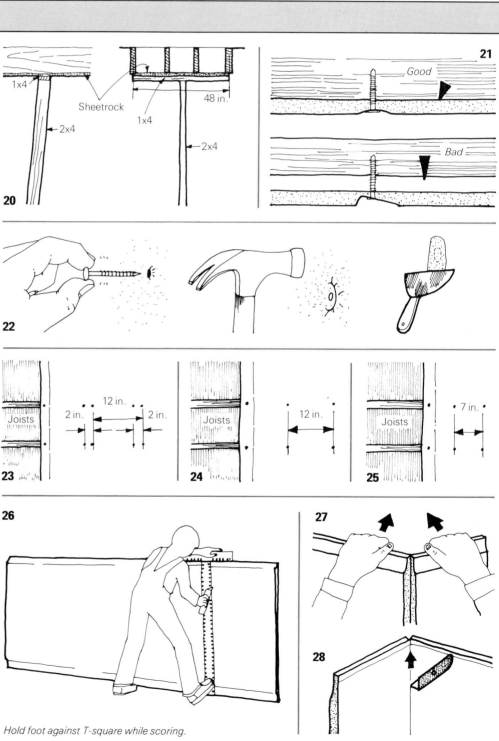

Hold foot against T-square while scoring.

plastic mud pan has the advantage of high sides, and runs about $4. At least four trowels are required—a 6-in., a 10-in., a 12-in. and a corner trowel **(17)**. The 12-in. trowel is much stiffer than the others, and it has a curved bottom **(18)**, so it takes a good bit of pressure to smooth and feather out the final coat. I use fiberglass tape for the ceiling joints, and paper tape for the corners and walls. Paper tape costs about $2 a roll, and a fiberglass roll **(19)** about $6. Fiberglass tape eliminates the first step of putting on a layer of compound. It has a sticky surface, and you just stick it on, and then apply a coat of compound—a big help when you're working overhead.

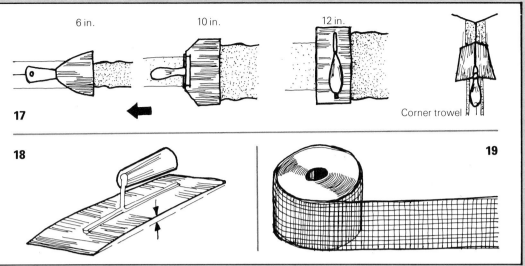

the first set of nails again as the sheet snugs up. Every nail should be driven just below the surface of the sheet, without tearing the paper, creating a dimple for the joint compound. Screwing is 12 in. o.c. **(24)**. Single nailing is 7 in. o.c. **(25)**. Cutting wallboard is probably the easiest part of the job. Just line up the T-square on the sheet and score **(26)**. One stroke with a sharp knife is usually enough. Snap it by folding the sheet back quickly **(27)**. Then cut along the backside **(28)**. If the edge needs cleaning, use the Surform.

Drywall walls are really simple. The panel that touches the ceiling should go in first to ensure a good joint there **(29)**. A few nails stuck into the sheet, on line with a stud, will make it easy to hold the sheet up snug and nail at the same time **(30)**. Leave about a ½-in. gap at the floor on the bottom sheet, so that a lever can be slipped under to push it up tight **(31)**. The bottom of the bottom sheet will be the cut edge, a good surface for the baseboard to butt against **(32)**. If you use corner clips or blocks at inside corners, install the sheet that runs parallel to the first so you don't have to nail into the blocks. The adjacent sheet is nailed to the stud in the corner and holds the first sheet in place **(33)**. Tape and spackle the whole business in place. The thing to remember about outside corners is that the drywall must extend far enough so that the corner bead has adequate backing **(34)**. Some professional drywallers cover the wall and then cut the outlet and switch holes. A safer way is to rub the outlet box with block chalk, hold the sheet in place, smack it with an open palm, and cut along the resulting lines left on the back of the sheet. Another way is to measure the location of the box and mark it on the sheet. This way isn't bad, but expect a few mistakes. Don't forget to mark all the openings, like medicine cabinets, recessed shelving and fans, as the sheets go up. They are easily lost.

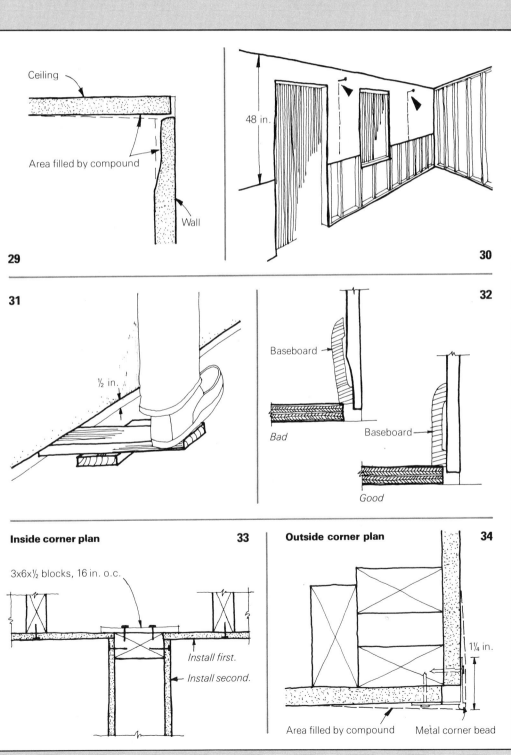

Spackling

Now we come to the spackling, and it's not as hard as most people think. With the proper technique and a little care, anyone can do a professional job. Cleanliness is a must. Any hard lumps or pieces of dirt in the spackling compound will mess up a joint (35). Load the hawk by scooping compound with a wood shingle or a 3-in. by 16-in. stick. Keep the scoop in the bucket, and the cover on so the compound stays moist and clean. Scrape the trowel clean as you work (36). If the scrapings are soft and clean, mix them with the stuff on the hawk. Try to keep the compound in a single gob so it won't dry out so fast. If it gets dirty or lumpy, dump it out. To do a joint, start with a 6-in. or 8-in. trowel and run a layer of compound as wide as the tape, the whole length of the joint (37). Holding the trowel at a low angle will leave a nice bed of compound (38). Lay the tape on the buttered joint and push it flat with the 6-in. trowel (39). Get all the bubbles out from under the tape. Spread a thin layer of compound over the tape. Keep it smooth; there's less sanding that way. Raise the trowel towards the perpendicular for smoothing (40). Try different angles until you get the best results. Each coat must be sanded when it's dry with 80-grit sandpaper. Silicon carbide works the best. The next coat is done the same way, but with the 10-in. trowel (41): Lay on a coat, wipe the trowel clean, then run it the full length of the joint to smooth it out. Press hard. If the trowel loads up, scrape the excess compound on the hawk and continue the run down the joint (42). Every time the trowel is picked up from the joint, it pulls some compound with it, leaving a ridge (43). Further, every bump

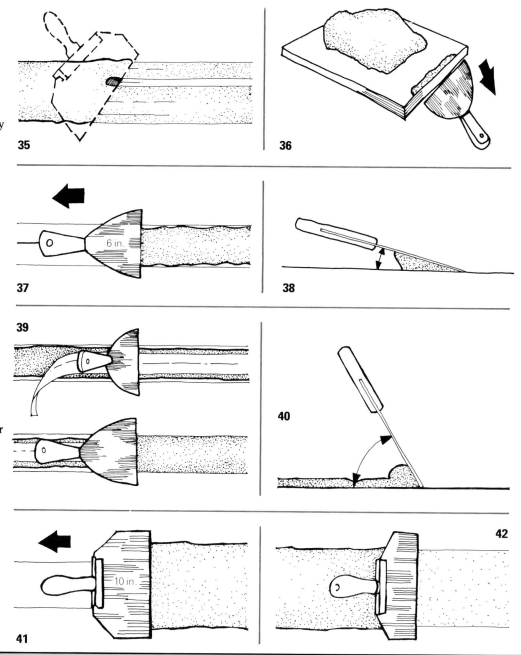

Finishing

Use the drying compound still on the hawk to fill the nail holes and dents. Run the trowel at a low angle to leave a layer of compound in the dimple (48). Then hold the trowel almost perpendicular to scrape the surface clean (49). Go to the next dimple; deposit, scrape, and so on. It will take a few coats, sanding between each coat. Always wear a mask when sanding. There is no asbestos in the compound, but breathing all that dust is not too healthy. Eye goggles are helpful, but you have to keep wiping them off.

I make a pad of sandpaper by cutting a sheet in half (50). Drag the sandpaper across a corner to break the glue (51). This will make the sandpaper more supple, and will prevent its forming a sharp ridge that will dig into the sanded area. Fold one

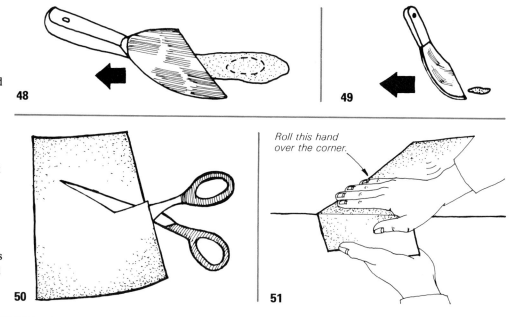

Roll this hand over the corner.

the trowel hits telegraphs onto the surface of the joint. Smoothing out may take a few strokes to get the excess off. Make that last stroke a nice long one, since it forms the base for the next coat. The last coat is done the same way, but with the 12-in. trowel (44). Try to make a long smooth run with this coat. If there's a long ceiling joint, it's best to have a long unobstructed run on a plank raised high enough to allow some good arm pressure against the joint. The 12-in. trowel is much stiffer than the others, so it takes some real force to smooth and feather out the final coat (45). This is the last coat, so make it a good one, and sand it with 120-grit paper. The better the joints are feathered, the less sanding you'll have to do. Sanding is dusty work, and I try to do as little as possible.

Inside corners are done in a similar fashion, but here the paper tape must be prefolded to fit the corner. Drywall tape has a crease down the middle, so folding is easy. The procedure is the same, but use the corner trowel to lay in the first coat of compound (46). Lay in the tape and smooth out the bubbles. A coat of compound is next; when it dries, sand it and then smooth on a final coat of compound. Experiment with the trowel angle for the best results. A few tries and you'll have it. Use a flat 6-in. trowel to scrape off the compound that squeezes past the edge of the corner trowel. Don't disturb the corner you just troweled.

Exterior corners are the easiest of all. Nail on the corner bead and spackle it with a 6-ir.. trowel (47). Do the last coat with the 10-in. trowel. There will be excess compound running around the corner as you trowel, but it's easy to scrape off because the corner bead gives you an edge surface to trowel against.

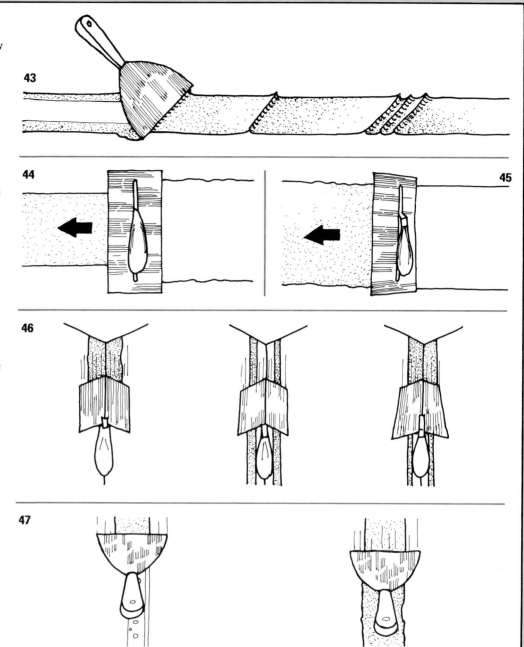

sheet into thirds (52), and then fold the other sheet in with it. This makes a nice soft pad. Sand using your fingertips (53), and stroke perpendicular to your fingers. The control is better, and the sanding surface is broader. Especially on ceilings and in corners, a pole sander (54) can save a lot of stretching and bending. It's worth the $12 purchase price. Sandpaper is easy to change, and the swivel head allows flat sanding from any angle of approach.

Check for spots you missed in sanding or spackling, and after you have admired the job, paint it. It's easy to paint without the trim in the way. Use a sealer first, or tiny filaments of paper from the sanding will show through successive coats. Pros spray-paint, and if the polyethylene vapor barrier is uncut and left over the windows, spraying is a breeze. □

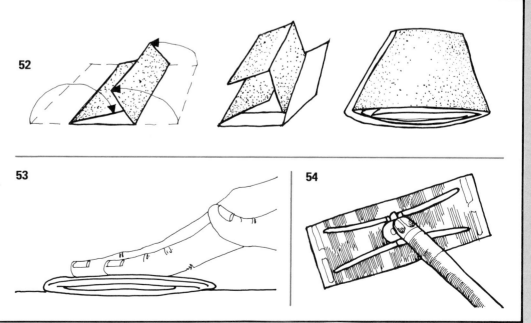

Two-Coat Plaster

How one builder avoids the drywall doldrums

by James Servais

I've spent much of my career as a contractor perfecting the flawless drywall finish. But a few years ago I was given the chance to try a rough style of plastering. Since then I've rarely applied the slick, smooth drywall. Now I prefer plaster. Its sensuous contours and the ease with which one can achieve curved and textured surfaces simply made the idea of angular drywalled rooms unappealing to me.

I use a two-coat method—one undercoat and a finish coat. I use neither screeds (lengths of angle iron pulled across wet plaster to level it) nor finish putty coats. The resulting walls are slightly irregular, wavy and textured, and as a consequence they take far less time and skill to apply than traditional plaster. This plastering works well with heavy timber framing and exposed wood ceilings, like those in Tudor or Spanish Colonial Revival homes. Obviously the rougher look won't work with every architectural style, but in the right context it adds a degree of authenticity that is hard to achieve any other way, and its rock-solid feel contributes to a building's sense of mass and permanence.

The substrate—About the only wood lath that you're likely to see these days is poking out of debris bins in front of older homes undergoing renovation. Contemporary plaster substrates are either expanded metal lath or rock lath. I use rock lath for my two-coat work. It is similar to gypboard—compressed gypsum that's covered on both sides by paper—but the sheets are smaller. They are ⅜ in. thick, 16 in. wide and 48 in. long, and they come in bundles that are easy to handle. Plaster sticks to this substrate because the multi-ply paper that covers the rock lath is very porous on the outside, causing a capillary action that makes the plaster adhere while it dries. The inner plies are water resistant, which keeps the plaster from drying out too fast.

Rock lath cuts like drywall and goes up fast. It's easy to deal with cutouts for switches, receptacles, light fixtures and beam notches because the pieces are small and maneuverable. Rock lath can be attached with drywall nails, drywall screws or (my favorite) stapled with a pneumatic staple gun (photo above right) loaded with 1-in. wide by 1½-in. long staples. Each 16-in. width of rock lath should have at least three

James Servais is a contractor in Berkeley, Calif. He teaches a course in two-coat plaster work for the Owner Builder Center (1516 Fifth St., Berkeley, Calif. 94710).

equally spaced fasteners into the framing—I prefer four. Unlike drywall, there's no advantage to neat work—just make it quick and firm.

Despite the recent controversy over the use of polyethylene sheet as an air-infiltration barrier, I use it over the insulation but under the rock lath. Once we have set the electric boxes to accommodate a ¾-in. thick wall, we stretch the plastic over the wall, boxes and all. We make a tight-fitting cutout in the rock lath to go around each box, which stretches the plastic pretty tight. The plastic stays there until the wall is plastered and painted. Then it's razored out, leaving electric boxes that aren't clogged with dried lumps of plaster. We also leave a couple of feet of plastic on the floor to help catch errant gobs of plaster.

Corners and curves—I learned how to do corners from a mad Rumanian, Dumitru Lo Bont. He told me that in his country you needed a whiskey bottle to make the corners. I bought him a bottle and he put a lot of plaster in the square corner, grasped the bottle by the neck and started to pull it down the corner. About a foot down he declared that the weight was incorrect and stopped to adjust the bottle. He started again and stopped, saying that the weight was not correct. He did this at every corner, until he came to me saying the bottle was now too light and he needed a new one.

Since working with Dumitru, I have discovered that you can adjust the radius of corners by putting up a narrow strip of rock lath at 45° across the corner, followed by wider strips of 3.5 diamond mesh (photo facing page, left). The mesh reinforces the plaster at these intersections, reducing the chances that the plaster will crack. The mesh should be attached to the framing on 6-in. centers with screws, staples or drywall nails.

Metal lath is good reinforcement in heavy traffic areas for outside corners that are liable to suffer the occasional bump. It also allows you the freedom to make curved shapes and soft reveals wherever the spirit dictates. For instance, one job I did had a low wall along the upstairs bedroom that was simply a 2x4 stud wall topped with a 2x6. Wrapped with lath and a built-up plaster step, it became a corbeled railing (photos facing page, right). Metal lath will provide you with the latitude to create unusual shapes, but remember that thick buildups over metal lath may require three or four coats of plaster.

To cut the metal lath, I use heavy shears or a skillsaw with a metal-cutting blade. If you de-

Rock lath is the substrate for today's plaster work, and it comes in 16-in. by 48-in. panels that are ⅜ in. thick. Panels are loosely butted together and fastened to the framing with nails, screws or pneumatically driven staples.

cide to use shears, be sure to wear a pair of leather gloves—shear-cut metal lath has sharp, serrated edges that will lacerate skin at the slightest touch.

The plaster—I've used two kinds of gypsum-based plaster to achieve the two-coat system described here: one kind uses sand as aggregate, and the other uses the lightweight perlite. Both U.S. Gypsum (101 S. Wacker Dr., Chicago, Ill., 60606) and National Gypsum (2001 Rexford Rd., Charlotte, N. C. 28211) make these products. The basic plaster is called either Dual Purpose Hardwall (U.S. Gypsum), or Two-Way Hardwall (National Gypsum). These basic plasters have to be mixed with #2 sand. U.S. Gypsum's Structo-Lite plaster has perlite aggregate, while National Gypsum calls their version Gypsolite. These plasters are premixed and need no further aggregate.

Once applied to the wall and finish-troweled, the two types have very similar textures. The Structo-Lite and Gypsolite weigh a little less than sand plaster, and are consequently a bit less grueling to apply. The sand plaster has greater mass, which is an advantage in passive-solar homes. In our 1,800-sq. ft. house, we placed approximately 22,000 lb. of sand-based plaster at about .36 Btu/lb. heat-storage capacity. If a heat sink is an issue in a house you are building, the

From *Fine Homebuilding* magazine (February 1986) 31:58-62

Radiused inside corners can be built up by nailing a narrow strip of rock lath into the corner, followed by a strip of metal lath. Here the author spreads the first coat of plaster across the wall to a depth of about ¼ in. In the upper right-hand corner of the photo, a halved terra-cotta flower pot has been wired to the wall. It conceals a porcelain light fixture. The pot will eventually be plastered to appear as though it's part of the wall.

Metal lath can be used to reinforce outside corners and to anchor plaster to wood armatures. At top, the lath wraps around a low wall and over its 2x6 cap. From the other side (above), the wall begins to take on its finished shape after the first coat of plaster has been applied.

Mixing plaster is a messy business, and adjacent structures or plants have to be protected with plastic sheeting. Once the plaster reaches the consistency of cake frosting, as shown above, it should be mixed for another fifteen minutes before it's applied to the wall.

After mixing, the plaster is transferred to a work stand and a load is scooped onto a plasterer's hawk.

When the first coat begins to set up, it is worked with a notched trowel (left) to create shallow grooves, which provide a mechanical lock for the second coat of plaster.

The second coat is finished with a steel trowel (above) that is cleaned frequently in a fresh bucket of water. Here the random texture of bumps, craters and swirls is beginning to show.

extra mass that comes along with a plastered interior may help reduce the need for other heat-storage systems.

Mixing—The best mixer for a residential job is a paddle-type plaster or stucco mixer. But to use this tool efficiently you have to have a lot of walls that need plaster and a crew ready to move a lot of mud. On small jobs I use an ordinary cement mixer (photo facing page, top left) and a hoe. The problem with the latter is that you can get lumps in the mix that can show in the finished wall, so you need to keep a close watch on the stuff as it mixes, breaking up any lumps you find with the hoe.

Mixing plaster is a messy job, so I protect the immediate area around the mixer with plastic sheeting. I dig a shallow basin outside the house, line it with plastic, and then place the mixer in the center. This helps to contain the inevitable spills that occur as the plaster is mixed and transported.

To mix a batch, put a few gallons of clean water into the mixer, turn it on, and then add your plaster mix until the mass looks like cake frosting. If you are using sand aggregate, your mix should be three parts sand to one part plaster. Let it mix five minutes, then pull out a handful and squeeze it. Water should come easily to the surface of the plaster, but the mixture shouldn't be soupy. Adjust water or plaster until you've got the right blend, and run the mixer about fifteen more minutes. It will take a few tries before you get it right—usually people mix it too dry at first. Once the plaster is blended, you have about an hour to work it.

On the wall—I pour the blended plaster into a 5-gal. bucket and carry it to my work stand, either a folding mortar stand or a pair of sawhorses with a plywood top. Then I pour the contents of the bucket onto the center of the table and scoop a load onto my hawk (a 1-ft. sq. piece of steel attached to a wooden handle) with a trowel (photo facing page, top right). I use either a 3x12 or 4x12 steel trowel with a wooden handle. I insist on wood-handled trowels—they feel good in my hand and they take a lot of abuse. These tools are readily available, but if you can't find them locally, you might try Marshalltown Trowel Co. (Box 738, Marshalltown, Iowa 50158) or Goldblatt (511 Osage, Kansas City, Kan. 66110).

If you're new to plastering, start work in a closet or room that won't be scrutinized too closely. Lean the hawk waist-high against the wall, and use the trowel to push about one quarter of the plaster onto the wall. Push the plaster upward in a large arc, move the hawk right or left and repeat the process until the hawk is empty. Once you've got the hawkful of plaster on the wall, go back and evenly spread the plaster until it is about ¼ in. thick. Don't try to make it perfect. When you have an even coat on one wall, keep an eye on it as it dries.

At the correct time—one to six hours, depending on the mix, temperature and humidity—go back and make horizontal scratches over the wall using a V-notched trowel. This makes a mechanical key for the next coat. Don't go too

deep. You shouldn't see any rock lath in the scratches. After a day or two this layer of plaster, called the scratch coat, will turn an even light grey color, and it will be crazed with little cracks. It is now ready for the second coat.

The next coat is the same material applied in the same way, spread thick enough to fill the scratches in the first layer (photo facing page, bottom left). If there's not much humidity in the air, lightly mist the scratch coat with water before applying the second layer of plaster.

Once you've got the second coat in place, wait until it just begins to set. Now get your steel trowel and a bucket of clean water and start smoothing out the finish surface to the desired texture. Dip the trowel in the bucket, then pull it across the plaster at a very slight angle (photo facing page, bottom right). You have to apply a fair amount of pressure to affect the plaster, so lean into it. If you find that you're moving the entire thickness of the second coat, you need to wait longer for the plaster to set. The more passes you make with the trowel and the more pressure you apply, the smoother the finished surface will be—just like working concrete. On curved areas where the radius is too tight for the 4x12 trowel, use a margin trowel (photo above).

I have found that many of my students get frustrated at this part of the job because they want to make the walls look smooth and unblem-

A margin trowel is useful for working the plaster on rounded corners (top). When creating the plaster roll that acts as a transition from the window stop to the wall plane (above), there's no need to worry about getting wet plaster on the wood if it has been sealed first. Smooth the edge with your finger and wipe the excess plaster off the trim with a damp rag.

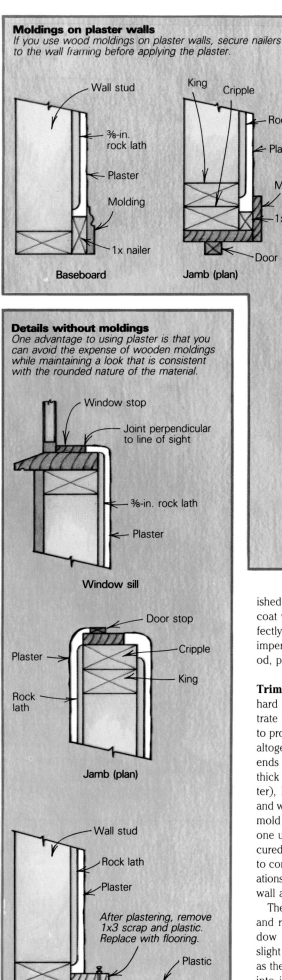

Moldings on plaster walls
If you use wood moldings on plaster walls, secure nailers to the wall framing before applying the plaster.

Wall stud
⅜-in. rock lath
Plaster
Molding
1x nailer

Baseboard

King
Cripple
Rock lath
Plaster
Molding
1x nailer
Door stop

Jamb (plan)

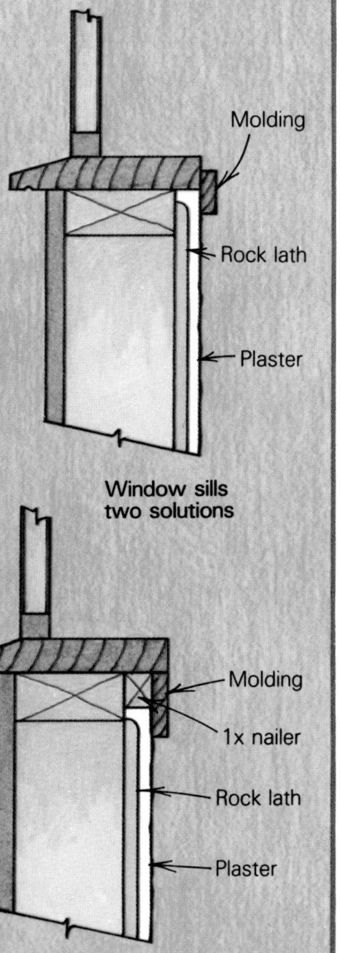

Details without moldings
One advantage to using plaster is that you can avoid the expense of wooden moldings while maintaining a look that is consistent with the rounded nature of the material.

Window stop
Joint perpendicular to line of sight
⅜-in. rock lath
Plaster

Window sill

Door stop
Cripple
King
Plaster
Rock lath

Jamb (plan)

Wall stud
Rock lath
Plaster
After plastering, remove 1x3 scrap and plastic. Replace with flooring.
Plastic

Trimless base

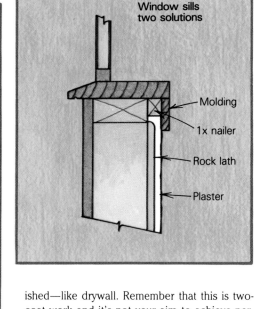

Molding
Rock lath
Plaster

Window sills two solutions

Molding
1x nailer
Rock lath
Plaster

rolls around the corners into window stops and door jambs (middle drawings, left). This puts the separation line at right angles to your view (you walk or look past them). To do this rolled effect, door jambs and window stops should be sealed before plastering. I use a margin trowel to place the basic roll of plaster (bottom photo, previous page), followed by a pass with my index finger to straighten the edge. If the woodwork is properly sealed, any plaster residue will come away with a wet rag.

In my own house we used no molding. Instead of a baseboard, the flooring meets the wall at a crisp edge. To make a slot for the flooring, I first attached scrap 1x to the floor after the plastic vapor barrier was in place, but before the rock lath went up (drawing, bottom left). After plastering, the scrap was carefully removed, leaving a slot for the ¾-in. oak flooring.

I ran the walls right up to our wood ceiling. I either protect exposed woodwork with masking tape, or take extra care to keep the plaster off unfinished wood surfaces. In most of the houses that I build, I sandblast the exposed woodwork to bring out the grain. Not surprisingly, this removes any traces of plaster on the wood. You can get the same results with a wire brush.

Imperfections in the finish can be scraped with a Stanley scraper or sanded and spackled. Wait two weeks for a plaster job to dry out, prime it with a good oil-base primer/sealer and then paint it as you would drywall.

Details such as light fixtures, built-in furniture and bookcases are all possible with plaster. The shape is framed and covered with lath or mesh, and then plastered. The same shapes would be very difficult in drywall.

My wife Gillian and I had just about run out of money when we were building our house, and needed some effective but inexpensive light fixtures. Enter the flowerpot sconce. Gillian had some empty hemispherical clay pots lying about. We sawed one in half and secured it with metal lath over a porcelain socket (photo, p. 97). With a couple of layers of plaster over it, the lamp appears to grow out of the wall, and it casts a warm indirect light toward the wood ceiling.

Plaster's drawbacks—Now for the bad part. Plastering is hard work. Spreading thick, cementitious goo onto walls and ceilings for hours on end is physically taxing. Arms, back, legs—everything will hurt, until you get used to it. The decline of plastering in the building trades is probably due in part to the fact that it is labor intensive, hence costly, and in part because few people want to work that hard anymore. Don't overestimate your strength on your first job.

The other consideration is cost. Because of limited production, rock lath remains expensive at about $.20 per sq. ft. The plaster comes to about $.40 per sq. ft. Considering the added labor expense, I estimate a plaster job at twice the cost of ½-in. skip-troweled drywall. On the plus side, if you can eliminate moldings, plaster gets closer, and if you have many curved walls or odd details it gets even more competitive. But if you love the beauty and freedom it gives you, plastered walls become a bargain and an attractive alternative to drywall. □

ished—like drywall. Remember that this is two-coat work and it's not your aim to achieve perfectly smooth walls. The uneven texture and imperfections are part of the charm of this method, part of what you're working toward.

Trim details—A layer of high-quality plaster is hard and brittle, and it's very difficult to penetrate with a nail. Consequently, you either have to provide nailers for moldings or eliminate them altogether. Since the finished layer of plaster ends up fluctuating between ¾ in. and ⅞ in. thick (⅜-in. rock lath plus about ⅜ in. of plaster), I use 1x nailers for baseboards and door and window casings (drawing, above). If a crown mold is called for, I use a nailer similar to the one used for the baseboard. The nailers are secured to the framing, and they work as screeds to control the thickness of the wall where fluctuations might otherwise show up as gaps between wall and trim.

The other option is to eliminate the molding and run the plaster directly into door and window jambs. The problem with this method is slight cracking at the wood/plaster intersection as the wood shrinks. Once the wood has settled into its eventual dimension, the cracks can be patched with Spackle and paint.

I like the way traditional Southwest architecture deals with this problem. There the plaster

Drawings: Frances Ashforth

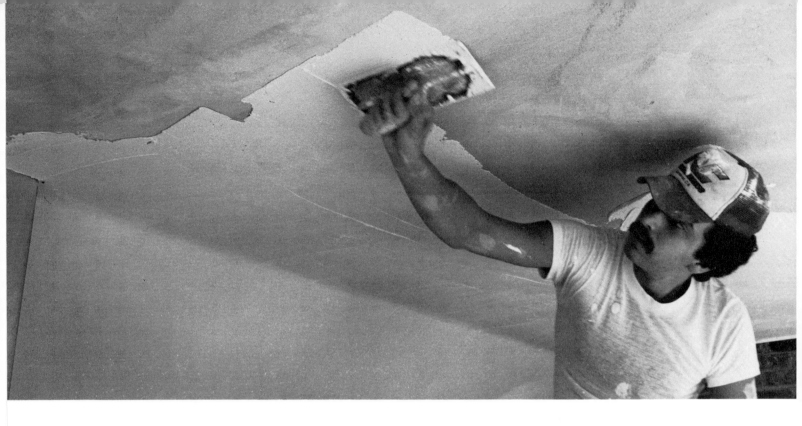

Veneer Plaster

Getting the look and texture of plaster on lath with gypboard and mud

by Tim Snyder

Finishing with plaster isn't the awesome job that it once was. Modern plastering systems have replaced the traditional wood or metal lath with gypboard sheets, textured on one side so they hold the plaster. After hanging and taping this substrate, you trowel on a thin (³⁄₃₂-in.), seamless surface that's far stronger (up to 3,000 psi) and harder than the drywall itself. It can be painted or left as is, with no sanding or other treatment required.

Called Thincoat, Skimcoat, Kalcoat, Imperial Plaster, or simply veneer plaster, depending on what manufacturer or contractor you're talking to, the finish costs a few cents more per square foot than a conventional drywall finish. It takes time to learn how to handle the mud, though, and you'll need at least two people to do the job right. The plaster sets quickly, and it's meant to be worked fast.

Materials—Veneer plaster isn't a new material, but it's only in the last several years that manufacturers have put together complete systems based on its use. The systems consist of gypsum-core backing board (which is nothing more than regular gypboard with a bluish, textured paper surface), more commonly called blueboard; high-strength plasters for one or two-coat finishes; a retarder compound to extend the setting time if you're mixing big batches or working slowly; fiberglass-mesh

joint tape; metal or plastic corner beads; and plastic edge terminals for transitions between plaster and other surfaces. You'll need all these items for most jobs. Manufacturers have their own product lines and encourage you to use their stuff only, but in practice, everything but the plaster pre-mix is interchangeable.

Veneer plaster can be either a one or a two-coat finish. The one-coat system is quicker, but if you want a slightly stronger, smoother wall surface, use the two-coat system. With the two-coat system, the base, or scratch coat is a grey, coarse plaster that bonds well with the substrate and with the white finish coat that covers it. This final coat can be floated to a satiny-smooth finish, or you can texture it by adding washed sand to the mix, or by going over the plaster roughly with your float. The finish coat can also be tinted with powdered pigment, though this is risky, since color may vary slightly from batch to batch.

Apart from sand, pigment and retarder, all you add to the plaster is clean water. The retarder gives about 15 minutes more working time—a blessing if you're troweling in corners or around contoured areas that demand more attention than flat work. A retarder works best when it's pre-mixed in water and then added to the batch as it's being mixed. Use the water-to-weight ratios recommended on the bag, both for plaster and retarder.

From *Fine Homebuilding* magazine (June 1983) 15:72-74

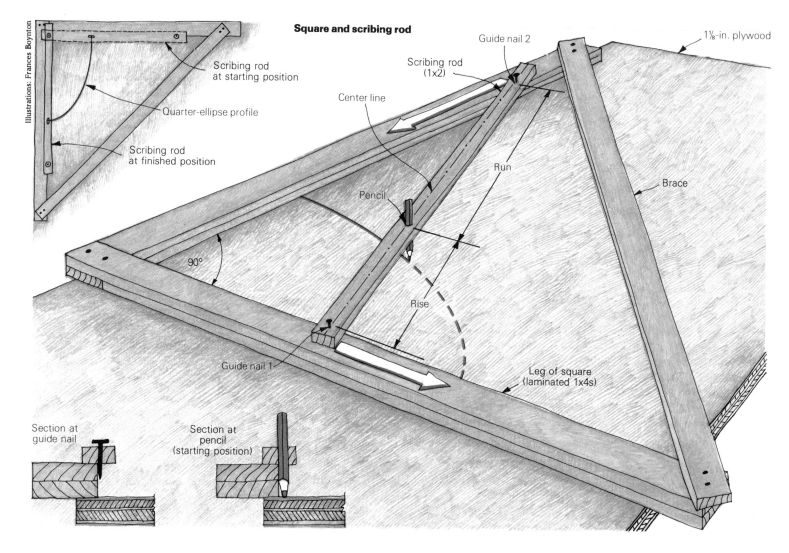

Square and scribing rod

Scribing rod at starting position

Quarter-ellipse profile

Scribing rod at finished position

Illustrations: Frances Boynton

Guide nail 2

Scribing rod (1x2)

Center line

1⅛-in. plywood

Run

Pencil

Brace

90°

Rise

Guide nail 1

Leg of square (laminated 1x4s)

Section at guide nail

Section at pencil (starting position)

The Scribed Ellipse
An easy way to lay out elliptical coves

by Jud Peake

A building with graceful curves has its own special signature. Soft transitions, impossible in angular structures, can be created with vaulted passageways and rooms with coved ceilings. For curved details, architects and builders usually choose the easily drawn radiused curve, or section of a circle. But sometimes the radius is aesthetically inappropriate, or just won't describe the curve needed at certain junctions. In cases like these, the ellipse may be a better shape. But unlike the radius with its equal rise and run, the ellipse is challenging to draw accurately.

An ellipse is a curve that has an equal combined distance from two centers (foci) at every point. The length and width of the ellipse are determined by perpendicular major and minor axes that cross at their centers. The foci are always located on the major axis, and are equidistant from its ends.

A common method for drawing the ellipse is to tie a loop of string around three pins, one at each focus and the third at one end of the minor axis (drawing, facing page, center). Once the string is in place, the pin at the minor axis is removed and replaced with a pencil. Exerting even tension on the string, draw the pencil around the foci and back to the starting point. This method, though, is not a very practical way to lay out work on a job site. Fortunately, there is an expedient carpenter's method for drawing the quarter-ellipse. All you need is a large square and a scribing rod.

Make the square out of two layers of 1x4s, lapping them at the right angle to keep the legs flush (drawing, above). Cut the legs about a foot longer than the combined rise and run of your quarter-ellipse. The hypotenuse, actually a brace to ensure an accurate 90° angle, can be lapped over the legs and nailed in place.

The scribing rod is the key to this operation, and carries the measurements and scribing tool (usually a carpenter's pencil) to transfer the quarter-ellipse profile onto the work. The rod should be about the same length as one leg of the square, and cut from a straight 1x2. A few inches from one end, drill a small pilot hole on the center line for an 8d nail, and drive the nail until its point protrudes about 1 in. from the bottom. The nail will act to guide the rod along the inside edge of the square. Measure the distance of the rise from the nail to a point on the center line and drill a hole for a pencil. Make sure you get a snug fit so the pencil won't

From *Fine Homebuilding* magazine (April 1982) 8:50-51

wobble. Continuing down the rod, measure the distance of the run from the pencil and set another guide nail on center.

To use the rig, position the triangle over the work (I like 1⅛-in. plywood for cove backing) and register the nails of the scribing rod against the inside edge of one leg of the square. With both nails tight against the leg, move the rod, all the while holding one nail against one leg, the other nail against the adjacent leg. Both pins should eventually come to rest tightly against the adjacent leg. The pencil will describe the curve of your ellipse. For ellipses with different rises and runs, move the guide nails to the appropriate spots.

Even if you make radiused (quarter-round) ceiling coves, the elliptical scriber can help you lay out the elliptical corner brace where the two coves are mitered together. For example, let's say we have an 18-in. radiused cove at a square corner in the walls. We know that the rise will remain constant at 18 in., but at the corner, the run will take a diagonal path across the ceiling, making it longer than the radius. You can determine the length of the diagonal by drawing the run of the radiused cove on the floor below the corner to be mitered and then taking a direct measurement. Another way to find the diagonal is to multiply the run by the square root of 2, or 1.415. Set the scribing-rod guide nails for the 18-in. rise and the 25½-in. run, mark the plywood bracing material, and cut out the brace with a jigsaw.

The inside corner brace has to be double beveled to meet the intersecting horizontal lines of the perpendicular coves. This bevel is complicated because it ranges from two 45° bevels at the bottom of the intersection to flat (180°) at the ceiling. To shape this bevel, I tack the brace, curved edge up, to a sawhorse and hollow out the working edge with a freehand circular-saw cut. When I make this cut, I keep both hands on the saw, and I don't stand behind the line of cut. Don't even try this cut if you aren't comfortable with the tool. This is an inexact method, but close enough for a corner that will receive several layers of plaster. The important thing is to hollow out the edge so the plaster won't be too thin at the corner crease.

An outside corner brace (photo right) is the same elliptical shape as its inside counterpart, but the double bevel comes to a point rather than a hollow. I cut them out with a jigsaw, and taper the angles on both sides, from 45° at the bottom to flat at the top.

The elliptical cove section has further applications. I recently worked on a Victorian building that featured a large oblong skylight with radiused ends. The skylight was to perch atop a room of the same shape and be connected to the walls by radiused cove sections. Because of a mistake on the drawing board, the skylight didn't end up exactly concentric with the radiused wall below. Instead of repositioning the wall or replacing the skylight, we decided to use elliptical sections instead of arcs, each one slightly different in dimension. □

Jud Peake is a carpenter and a contractor. He lives in Oakland, Calif.

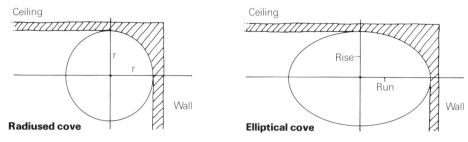

Radiused cove **Elliptical cove**

In section, the most common coved ceiling is a quarter circle, with equal rise and run. When the rise and run differ, the section is elliptical.

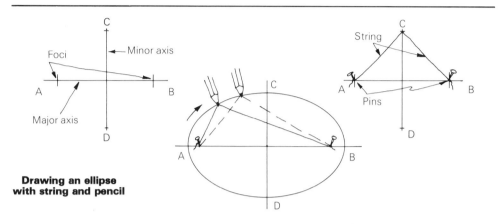

Drawing an ellipse with string and pencil

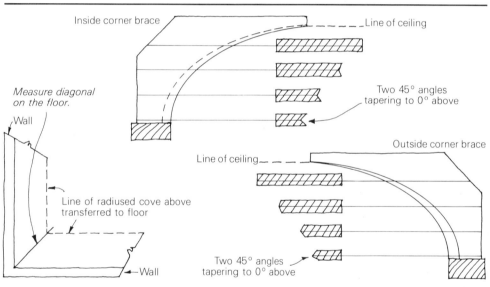

Elliptical corner braces *(below) for a radiused cove ceiling can be drawn with a square and scribing rod. Both inside and outside braces are double-beveled (above) to hold intersecting pieces of lath.*

Hardwood Strip Flooring
How to lay out and install it

by Don Bollinger

Putting down a hardwood floor is one of the few kinds of woodworking that can be classified as both rough and finish carpentry. Anyone who has installed a ¾-in. oak floor will attest to the brute force required to drive home the nails, and to the telltale evidence of every mistake. And since all the work happens at ground level, a floor mechanic spends half his time on his knees and the rest stooped over like a field hand.

Still, the bundles of hardwood flooring that get broken open for each job contain some of the most carefully graded and highest-quality building materials available today. It's a pleasure to work with this stuff, and when wood this good is properly installed and cared for, it makes for a durable floor that can add a lifetime of color and warmth to any home.

Strip flooring—There are two basic types of strip flooring: tongue-and-groove (T&G) and square-edged. Some T&G strips are end-matched (drawing, above right), which means they're T&G on their ends as well as along their lengths. Strip flooring ranges from ⁵⁄₁₆ in. to 1½ in. thick, and from ¾ in. to 3¼ in. wide. Here in the Northwest, most strip flooring is red oak, white oak or maple, and the most common size, as it is throughout the country, is ¾-in. thick by 2¼-in. wide T&G. I like end-matched strips because they give the floor a bit more strength, and reduce bouncing where strips butt end to end between joists.

Hardwood strip flooring is further subdivided into random lengths and shorts, depending on quality. The better the grade, the

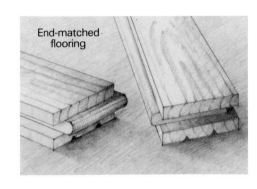
End-matched flooring

fewer the knots, sapwood and pinworm holes, and the longer the strips. For instance, the best grade of oak is called *clear*, and the strips are at least 1¼ ft. long, averaging about 3¾ ft. They are usually flatsawn, but the more attractive quartersawn strips can be special ordered. The poorest grade is *#2 common*. These strips, which average 2¼ ft. in length, have knots, checks and worm holes. When you order any grade, add 5% to 10% to allow for waste.

Precautions—Have your flooring delivered at least three days before you install it, and break the flooring bundles open so the boards can adjust to the ambient moisture. If the floor is going into a new house, make sure the place is sealed up, and that all masonry, plaster, drywall and other wet-process work is completely dry. Three weeks before the flooring is to be delivered, turn up the thermostat to your typical setting to help dry things out. If there is too much moisture in the house when

the floor is installed, the floorboards will shrink when the humidity drops, and you'll get ugly gaps between them. Moisture can also come up through the subfloor, so make sure the basement is dry and well ventilated.

Subfloors—A hardwood floor is no better than what's under it. Almost all squeaky floors, cracking finishes and other signs of early aging can be traced directly to inadequate underlayments. I've found it best to install a subfloor that's heavier than those generally specified in the building code (see the chart below). You will need an especially heavy subfloor if you're planning to lay your strip flooring parallel to the joists.

Plywood is the most common underlayment for strip flooring—I don't use particleboard because nails don't hold in it and water deteriorates it. For ¾-in. thick strips (like the ones we're talking about in this article), use ¾-in. plywood blocked under all edges, with a ⅛-in. expansion gap between each panel. Glue the plywood sheets down with panel adhesive and nail them with either hot-dipped galvanized or ring-shanked nails. This combination will reduce movement and squeaks.

If you use planks for your subfloor, they should be 6 in. to 8 in. wide and at least ¾ in. thick. They can be either square-edged or shiplapped, but shouldn't be T&G unless they are at least 1½ in. thick—anything less will flex and squeak.

Without a moisture meter, it's hard to tell whether boards will expand or contract. To play it safe, leave ⅛-in. gaps between your

Orientation of flooring	Random-length strip		Shorts (18 in. or less)	Square-edge strip
	up to 3¼ in. wide, ¾ in. thick	4 to 8 in. wide, ¾ in. thick	up to 3 in. wide, ¾ in. thick	1 to 3 in. wide, ½ to ⁵⁄₁₆ in. thick
Strips running at 90° to joists set 16 in. o.c.*	¾-in. T&G plywood ¾-in. square-edge plywood ¾-in. by 6 to 8-in. plank or shiplap laid diagonally or two layers of ½-in. square-edge plywood with staggered seams	⅞-in. T&G plywood 1-in. square-edge plywood ⅞-in. square-edge plywood ¾-in. by 6 to 8-in. plank or shiplap laid diagonally or two layers of ½-in. square-edge plywood with staggered seams	1-in. T&G plywood or square-edge plywood Two layers of ½-in. square-edge plywood with staggered seams	1⅛-in. T&G plywood or square-edge plywood Two layers of ⅝-in. square-edge plywood with staggered seams
Strips diagonal or with joists set 16 in. o.c.*	1⅛-in. T&G plywood, or two layers of ⅝-in. square-edge plywood with staggered seams Bridging between joists			
* For joists set 24 in. o.c., add ¼ in. underlayment; for joists set 12 in. o.c., subtract ⅛ in. underlayment.				

Table title: Recommended underlayments for tongue-and-groove hardwood flooring

subfloor planks. A standard ¾-in. plank sub-floor should be installed diagonally across joists spaced 12 in. to 16 in. o.c. Thicker T&G material may be laid at right angles to the joists. In either case, be sure that all underlayment planks butt end to end over a joist or blocking. Otherwise you'll get soft spots, fractured finishes and squeaks.

Whatever underlayment you use, be sure that its moisture content has stabilized before you install any hardwood over it. Don't use for underlayment plywood that has been used in concrete forms, because it has absorbed a lot of moisture and could delaminate.

Once the subfloor is installed, any differences in height between adjoining pieces of underlayment should be sanded flat with a belt sander using a coarse-grit belt. Scrape away any plaster or joint-compound lumps, then sweep or vacuum the subfloor. Prowl it one more time with a handful of 2-in. ring-shanked nails, and fix any humps and squeaks. Finally, mark the location of each joist at the base of the wall.

It's usually best to run the flooring in the direction of the longest dimension in the room. This will reduce the area over which the flooring will do most of its seasonal expansion and contraction. A vapor barrier between subfloor and finish flooring will keep out dust and moisture from below, help prevent squeaks and stop unwanted airflow through plank subfloors. I usually use three-ply resin paper, but 15-lb. to 30-lb. asphalt-saturated felt works just as well. Roll it out in the direction you want to run the strip flooring. If the floor is directly over a furnace or uninsulated heating ducts, add an extra layer of paper or felt to help keep the finish flooring from drying out.

Cover the length of the room where the floor-laying will begin with a few courses of VB paper or felt stapled to the floor. The baseline (see next page) is chalklined onto the felt. Subsequent courses needn't be stapled. Overlap the edges of the strips 3 in. to 6 in., and lay more down as the floor progresses, to keep traffic from tearing up the paper.

If the total thickness of the subfloor plus underlayment is ¾ in. or less, nail ¾-in. strip flooring into and between all joists. Snap chalklines between the marks on the wall so you know where the joists are. You'll have to restrike this line each time you lay a new strip of felt.

Moisture and movement—Like any wood, each piece of hardwood flooring expands across its grain as it picks up moisture, and contracts as it loses it. Given a large lateral area, the cumulative effect can cause a floor to

Installing a hardwood strip floor requires planning, especially in places like hallways, where misaligned boards are easy to see. Once the first few rows are nailed in place, the floor is racked, or loosely laid out, several rows at a time. When the flooring is extended into the kitchen (at right), a hardwood spline will connect the grooves of adjacent boards, allowing a change in the direction of the leading edge.

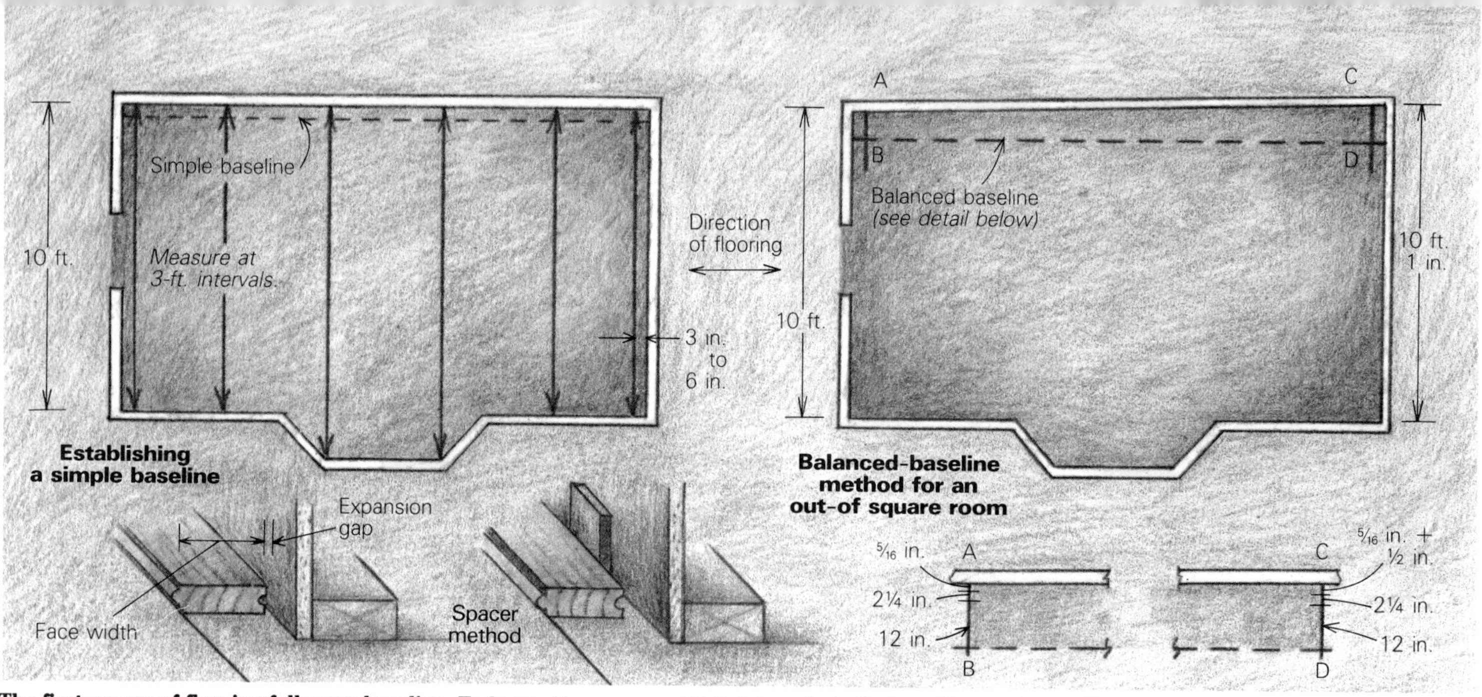

The first course of flooring follows a baseline. To locate it, measure at 3-ft. intervals to find out if opposing walls are parallel. If they are within ½ in., use a simple baseline, left. The balanced baseline, right, averages the error in a room whose opposing walls are more than ½ in. off parallel.

buckle. I've seen entire walls uprooted at the plate as a result of a leaky roof.

A little seasonal movement is normal. To allow for expansion, leave gaps between the edge of the floor and the wall. Since a baseboard or shoe molding will conceal this gap, many installers simply allow ¾ in. around the entire perimeter. A rule of thumb is to expect hardwood to expand and contract 1/16 in. for every running foot of cross-grain flooring.

Wood movement along the grain is negligible, but we leave about a ½-in. gap at each end wall. This space makes it easier to fit the boards together and allows air circulation.

Nail schedule

Flooring (in.)	Fasteners: Size	Spacing*
Tongue-and-groove flooring		
¾ x 1½ ¾ x 2¼ ¾ x 3¼	2-in. machine-driven 7d or 8d screw, or cut nail	10 to 12 in.
¾ x 3-in. to 8-in. plank**		8 in. into and between joists
½ x 1½ ½ x 2	1½-in. machine-driven fastener or 5d cut steel or wire casing nail	10 in.
⅜ x 1½ ⅜ x 2	1¼-in. machine-driven fastener or 4d bright wire casing nail	8 in.
Square-edge flooring		
5/16 x 1½	1-in., 15-ga. fully barbed flooring brad	2 nails 7 in. o.c.
5/16 x 2 5/16 x 1⅓	1-in., 15-ga. fully barbed flooring brad	5 in. o.c. on alternate sides of strip

*Dimensions in this chart are nominal, not actual, size. Flooring ½ in. or thinner must be laid on a subfloor. Tongue-and-groove flooring must be blind-nailed. Square-edge flooring is face-nailed through top face. *If the subfloor is ½-in. plywood, flooring should be nailed into each joist, with additional fastening between joists. **If planks are wider than 4 in., screw or glue them to the subfloor.*

Layout—Assume there is no such thing as a square room. I know I've never found one. The most carefully constructed rooms are usually out of square by ½ in. to ¾ in. These inaccuracies are easy to mask in the typical strip floor just by allowing for them in the expansion gaps around the edges of the room. Severe cases of out-of-square can call for a tapered floorboard hidden under a cabinet kickspace or a counter overhang, or a mask in the form of some type of border arrangement. In one terribly out-of-square room, I finally installed the flooring diagonally, after checking to make sure the subfloor would support it.

A good installer will spend as much as a third of the job sizing up the layout, installing the first few boards and anticipating and avoiding the glitch that would foul up the floor. Remember that some wall-to-floor intersections are in more visible locations than others, and that the baseboards and floorboards should be as close to parallel as possible. Converging floorboard and baseboard lines look especially awful in hallways. The error caused by out-of-square walls has to be dispersed gradually. Don't try to deal with it at a single intersection, where it will smack you in the eyes.

When you're installing strip flooring you need to worry only about the alignment of walls parallel to the flooring. Run a tape measure between these walls every three feet along their lengths. Start 3 in. to 6 in. away from the corners. They tend to flare out because of the way wallboard is taped. If there are more than two facing walls, make at least two measurements for each additional wall. If the opposing walls are less than ½ in. out of parallel, you can use a simple baseline to begin laying the floor.

Baselines—The baseline is a starting point for the flooring in a room, and is used as a reference device to keep straight, even runs of flooring. Since every installation is different, the location of the baseline is never quite the same, but there are some guidelines to determine where to put it. It usually parallels the longest wall or longest uninterrupted run of flooring in the room. Given a choice between two walls of equal length, establish the baseline near the most visible wall.

In the simplest situation, you go to the end of one wall and pencil a mark on the vapor barrier. The distance from the wall to the mark is equal to the face width of one floorboard plus the expansion gap, as shown in the drawing, above left. Repeat the procedure at the other end of the wall and snap a chalkline between the two marks. This is the baseline. The first course of flooring is then face-nailed in place with the baseline as a guide. Begin nailing halfway along the length of each board, then work toward each end.

If the opposing walls are close to parallel, an easier way to start the first strip is to place removable shims equal to the thickness of the expansion gap against the wall. The first strip butts these shims, and is nailed in place. If you are using this method, be sure to sight along the leading edge of the flooring to make sure any dips or wows (depressions or bulges) in the wall aren't being telegraphed into the strips.

The balanced baseline—If opposite walls are more than ½ in. out-of-parallel, the baseline has to average out the discrepancy and distribute the error to both sides of the room. This is called a balanced baseline.

In the drawing above right, one wall is 10 ft. long and the opposite wall is 10 ft. 1 in. Start at point A and measure out 14 9/16 in. to find point B. This dimension is the total of the expansion gap (5/16 in. per side for 10 ft. of cross-grain floor) plus the flooring face width (2¼ in.) plus a 1-ft. workspace. At point C, add ½ in. to the measurement to account for half of the 1-in. discrepancy. Mark point D. This total measurement is 15 1/16 in.

Now snap a chalkline between point B and point D, and forget about any other reference

Chart is adapted from the National Oak Flooring Manufacturers Association

line. Measure back 1 ft. to locate the leading edge of the first piece of flooring.

Sometimes a balanced baseline alone won't solve a special problem. We once had a job in which a stairwell, in a run of only 2 ft. 4 in., wandered ⅜ in. out of parallel. We decided to compensate for the error with two different steps. We added ½ in. to one end of the baseline measurement, bringing the stairwell and baseline a little closer to parallel, then we took up the rest of the error by slightly skewing the nosing around the stairwell. This left that opening slightly out of square, but no one without a tape will ever notice.

Installing the first boards—Sort through your longest bundles and find enough straight pieces for two starter and four finishing rows. Check them by sighting down the boards as though you were looking for a good pool cue. Using bent or crooked pieces to finish out a floor usually results in ugly gaps in the last few rows. It's hard enough to pull the flooring up tight next to a wall without having to fight recalcitrant boards.

For right-handers, it's easier to work from left to right, so most floors begin at the left-hand corner (drawing, top right). Set the first strip squarely on the baseline ½ in. out from the wall; the tongue side should face you, with the end-matched groove to your right.

Align the leading edge of the groove side with the baseline. Starting in the middle of the board's length, about ½ in. from the grooved edge, face-nail ¾-in. boards with 8d cut nails or 2-in. power-driven fasteners (the chart on the facing page gives other nailing specs).

We use a power nailer to set the boards (photo right). It fits over the leading edge of the board, and with one blow of the mallet to the nailer's plunger, the tool draws the board tight against its neighbor as it drives a barbed fastener through the top of the tongue and into the subfloor at a 45° angle. We use a power nailer designed for face-nailing in tight spots near walls.

Power nailers are fast and easy to use. They lessen the chance of dings from misguided hammer blows. You can rent one from a tool-rental yard or a flooring-supply center, or buy your own. Two companies that make them are Porta-Tools (Box 1257, Wilmington, N.C. 28402) and Power Nailer (Power Nail Co., Rte. #22, Prairie View, Ill. 60069).

Place the nails 8 in. o.c., and make sure every other nail sinks into a joist. At board ends, drill ¹⁄₁₆-in. pilot holes to keep the boards from splitting. Blind-nail the same number of fasteners through the board's tongue, taking care to avoid knocking the board out of alignment. Set all nails below the face of each board.

Select strips for the second row that won't end within 3 in. of the butt joints in the first row (drawing, center right). You should stagger butt joints this way throughout the floor. Tap the boards in place using a buffer

One blow to the power nailer is enough to pull the flooring tight against the previous course and set a barbed nail.

Illustrations: Frances Ashforth

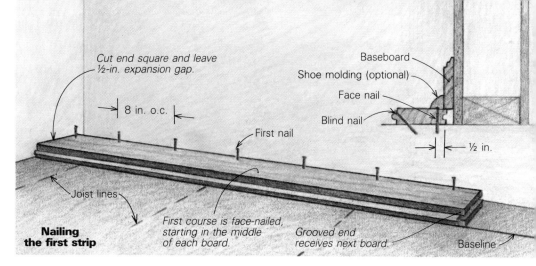

Cut end square and leave ½-in. expansion gap.

8 in. o.c.

First nail

Baseboard

Shoe molding (optional)

Face nail

Blind nail

½ in.

Joist lines

Nailing the first strip

First course is face-nailed, starting in the middle of each board.

Grooved end receives next board.

Baseline

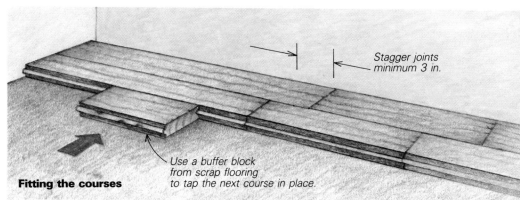

Stagger joints minimum 3 in.

Use a buffer block from scrap flooring to tap the next course in place.

Fitting the courses

Two header boards intersect in a mitered corner. The door casing has been trimmed off at the bottom to allow the flooring to slip under for a finished look. The tape measure registers the distance from the board's face edge to the baseline.

At a doorway, a header board is screwed in place 8 in. o.c. with #10 2-in. square-drive screws. Hardwood plugs, with their grain turned at 90° to the grain of the header board, will fill the counterbores. The first course of flooring will be mitered to meet the edge of the header board.

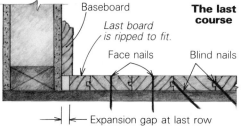

The last course

Baseboard

Last board is ripped to fit.

Face nails   Blind nails

Expansion gap at last row

The last course. Strips are levered into place, above, and face-nailed with a power nailer.

At row ends, left, boards are marked for length by eye and cut off at 90° with a power miter box. Trimming this board will remove a flaw, making it a good candidate for a row-ending piece. Another common practice is to pick end pieces that leave cutoffs at least 6 in. long, which are used to start subsequent courses.

board—a short offcut used to protect the board from hammer blows while positioning it—and blind-nail them in place. In most cases you'll get a snug fit with little effort, but occasionally a crooked piece will require levering with a prybar or a screwdriver. Before you exert too much force, check to see if a board is hung up on a cracked tongue, on an unset nail or on a small bit of wood.

Once you've gotten this far, you can move right along. I like to keep my power miter box just behind me and to my right as I'm nailing. It's easy to reach there, and the sawdust it makes is away from my clean vapor barrier.

Once the first few rows are nailed down, we "rack" the floor. This means laying out several loose rows of flooring a few inches ahead of the last row installed (photo p. 107). We place the pieces so the butt joints are as far apart as possible, and so that short pieces get evenly distributed. We use long strips across doorways and down halls, and save the shorts for closets, bathrooms and other confined areas.

Unless a precise cut is needed, we don't measure cutlines at the end of a row. Instead, we hold the board against the wall so that its tongued edge faces the nailed-down rows and eyeball a cutline (photo bottom left). When the board is drawn tight into its end-matched groove, there is a ½-in. gap at the wall.

If the flooring butts something that isn't already in place, such as a threshold, hearth or finished trim, we work to a temporary straightedge, start the flooring at it and face-nail the ends of each board.

Header boards—At doorways and passages, we install header boards for decorative effect. They're wider T&G planks than the rest of the flooring, and they're screwed to the subfloor rather than nailed (photos facing page, top). Instead of blending into the flooring with a butt joint, they join at a 45° angle. Their countersunk screw holes are plugged using the same wood, with the grain turned 90° for a subtle highlight.

You can't leave an expansion gap when you install header boards. Instead, we add 50% to the minimum gap at the opposite wall.

Check the alignment of the rows as they march across the room to make sure the occasional dip or wow isn't becoming amplified. Usually they can be handled with a buffer board and a small sledge, but sometimes a board will need a little wood removed. I use a plane to take material off the groove side of the strip to match fluctuations in neighboring boards. Cull out the really distorted pieces—they're just too much trouble to deal with.

When you reach the opposite wall, you'll need to face-nail the last few rows while they're held tightly in place. Don't do more than two rows at a time. Get out those straight boards you've been saving and use a piece of flooring held against a block on the wall to lever them into position (photo facing page, bottom right). Position the force of the lever over a wall stud, as shown, and use a block long enough to span two studs to keep from damaging the wallboard.

Once the last full row has been nailed down, you'll probably need to rip one last course to finish the floor (drawing, facing page). Occasionally a ripped piece will be too narrow to nail without splitting. In such a case, I glue the narrow piece to the last full strip with yellow glue and temporarily wedge it in place until the glue dries.

Stairwell nosing—When hallway flooring is nailed off, we take another set of measurements to see how close the strips and the stairwell are to being parallel (photo above right). If things are a bit off, we pencil-mark the necessary adjustment on the subfloor and adjust the nosing strips to compensate for the error. Then we temporarily screw them in place. When it's time to install the notched floorboard (photo below right), we remove the nosing so that we have room to tap the grooved edge of the flooring into place. If there's no nosing to adjust, we dress a slight taper into a number of strips with a plane to hide the error.

Finishing up—We wait two weeks before we go back to a floor to sand and finish it. Here in the humid Northwest, this allows the newly installed floor to acclimate itself to its new surroundings, and to adjust to the stresses brought on by being nailed and screwed to a flat surface. After sanding, we stain the floor and finish it. And that's another story. □

Don Bollinger owns the Oak Floors of Green-bank in Seattle, Wash.

The distance from the advancing floorboards to the stairwell nosing is measured to check for parallel, above. Once this is established, the nosing is cut to length and screwed in place. In the finished floor, below, the strip next to the angled nosing piece required notching. The nosing was removed to let in the strip, then replaced. The countersunk screw holes were plugged.

Floor Sanding

Follow a pattern with powerful machines for best results

by Don Bollinger

Sanding a floor is nasty work. It's noisy and dusty, and it takes a fair amount of practice to do it right. It's one of the most frequently subbed-out jobs. But a lot of people are willing to take it on for the first time, as owners of tool-rental shops will readily attest. Here are some tips, and a few precautions.

Preparation—Take everything that isn't nailed down out of the room. Cover the built-ins with plastic sheeting, and tack sheets of plastic across any doorways. Rolled-up towels laid against the bottom of the far side of doors in rooms being sanded will also help to contain the dust.

Sweep the floor clean and set any protruding nails at least 1/16 in. below the surface. Repair loose boards or squeaks with nails driven into joists. If there's room under the floor, I like to fix squeaks by driving screws through the subfloor to draw the hardwood tight against the underlayment. These screws have

Don Bollinger owns Oak Floors of Greenbank in Seattle, Wash.

to be 1/4 in. shy of the total thickness of the subfloor and the flooring.

You'll need a dust mask while you're sanding. You may want to wear some ear protection, too. Wear shoes that don't have crevices in the soles that can pick up grit. Sneakers or running shoes are good, but avoid the ones with black soles—they can leave scuff marks that are hard to remove.

The tools—It takes two basic kinds of sanders to finish (or refinish) a floor: a drum sander and a power edger. Both are available at tool-rental shops, and both require some muscle and practice to use correctly.

The drum sander, or floor sander, is used to sand most of the floor (photo facing page, top). It is a formidable machine. Even the smaller versions weigh about 125 lb., and they look like a cross between a lawn mower and a steamroller. They need their own 15-amp circuit to operate an 8-in. drum that rotates at about 5,000 rpm. When this drum is fitted with coarse-grit paper and lowered onto the floor, it wants to take off like a dragster. If you

hold it in one place, it is inclined to eat its way through the floor and into the basement. It's not a machine to be taken lightly.

But it's the only tool for the job, and with some practice, an operator can develop the required light touch. The first-time user should practice on a section of floor that won't be in direct view. Try a bedroom floor or part of a room that will be covered with a rug. And sand only with the grain.

As in any sanding job, you start with coarse-grit paper and work up to the fine grit. It takes quite a few sheets of at least three different grits to do an entire floor. Sheets are sold at the rental shops, and you can generally return the ones you don't use. Take plenty.

If you are refinishing a floor that's covered with paint, begin sanding with a very coarse paper—12 to 20 grit. For a new floor, start with 24 to 40 grit.

The drum sander is designed to make a slightly deeper cut on the left side of the drum (drawing facing page, top). This delicate angle allows you to feather the edge of the cut on the right side. To benefit from this feature, you should start sanding on the right side of the room, and work toward the left.

Begin about a third of the way up the floor (drawing facing page, center), and gradually lower the drum to the floor by letting up on the handle. It's important that you walk forward as you do this so the drum won't dig in in one spot. The weight of the machine will do the cutting. You want to make sure that the drum smoothly engages and disengages with the floor.

As you near the wall (about 1 ft. away), begin lifting up the drum, and then lower it again as you back up over the same path. You're towing the sander now, and this is when it does its best cutting.

Move to the left in 2-in. to 4-in. increments, making a forward and a backward pass over each section. When you've covered two-thirds of the floor, go to the left wall, turn around and sand the remaining third in the same manner. Take care to feather the slight ridge where you changed direction.

Sometimes a floor will be so uneven that it has to be sanded diagonally to the strips. Do this very carefully, and only with the coarsest paper. Start in one corner and move from right to left until two-thirds of the floor is covered. Go to the opposite corner, reverse direction and finish the remaining third. Then

sand the entire floor in the direction of the grain with the coarsest paper.

Since the drum sander can't reach in close to walls, corners and other tight areas, you'll have to sand these surfaces with an edger (photo bottom right). This powerful disc sander has grips for both hands built into the body. When it's tilted back on its wheels, the disc is lifted off the floor. When allowed to tilt forward, the machine begins its work. Like the drum sander, it is a difficult tool to use correctly without some practice. Try it out where you can't do too much damage.

The edger has a light mounted on the front of its chassis that helps you see what you're doing. If you're renting one, check to see that the bulb works—frequently they don't.

Unlike the drum sander, the edger makes its deepest cut on the right-hand side (at about one o'clock), and should be moved across the floor from left to right. A standard pattern for moving the tool is the semicircular path shown in the drawing, bottom right. There are other ways to operate an edger, too. Do whatever works best for you. Sand the areas missed by the drum sander with the same or a slightly finer grit.

When the entire floor has been sanded with coarse-grit paper, fill any holes in the floor. I use a lacquer-base filler blended to match the species of wood that I'm finishing, and I spot-fill nail holes and cracks between boards with a putty knife. If it's a top-nailed floor with a lot of nail holes or a parquet floor with numerous gaps, I trowel on the filler with a concrete trowel and wipe away the excess with a burlap rag. When the filler has dried, the floor can be sanded with medium (40 to 60-grit) paper. When you've finished with the drum sander and edger, check for shiners—nailheads turned silvery from being sanded. Set them, and refill the holes.

Final sanding should be done very carefully with 80 to 120-grit paper. Feathering is most important now because any ridges will show in the finished floor.

When the floor has been completely drum-sanded with the last paper, clean up the corners. For this task, the professionals I know use a common paint scraper, sharpened to a razor edge with a file or stone. Most of the time you should pull the scraper with the grain, but for hard-to-reach crannies you may have to work it at a 45° angle. This is acceptable practice, but never pull across the grain. When the scraping is done, hand-feather the perimeter with a sanding block wrapped with the final grit.

Cleanup—Sweep the ceiling, walls and floor as clean as you can with a good broom. Next, lightly dampen a medium-sized towel with paint thinner, lacquer thinner or alcohol, wrap it around the broomhead, and go over the entire floor with it. This is called tacking the floor, and it will collect most of the fine dust that still remains. Thoroughly vacuum the edges and corners using a crevice attachment, and you're ready to apply stains, sealers and finishes. □

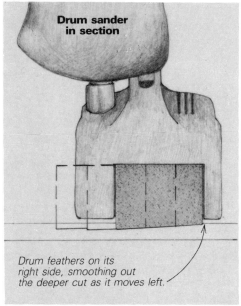

Drum feathers on its right side, smoothing out the deeper cut as it moves left.

Drum sander in section

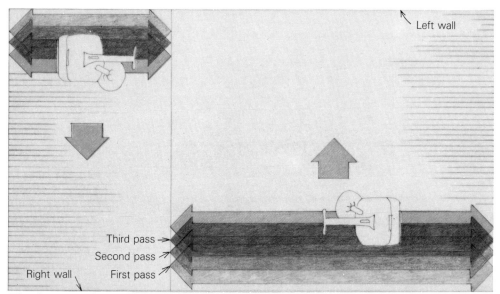

Left wall

Third pass →
Second pass →
First pass →
Right wall

The drum sander, top left, is heavy and unwieldy. Using one takes some getting used to, so if you'll be doing your own floor, practice in an area that will be out of sight. When you're ready, begin sanding a third of the distance up the floor, as shown in the drawing, above. Push and pull the sander over the same area before advancing 2 in. to 4 in. for the next pass. When you've done two-thirds of the floor, reverse direction and sand the rest of the floor in the same manner.

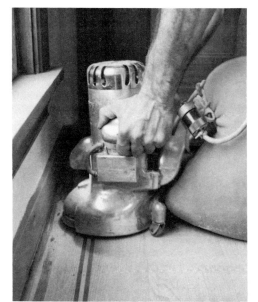

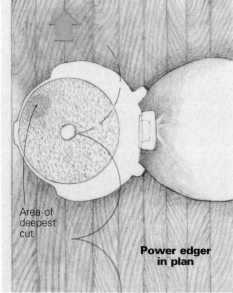

Area of deepest cut

Power edger in plan

The power edger sands the spots you can't reach with the drum sander. Its heavy metal guard keeps the disc from touching the baseboard. A looping, semicircular motion, as shown at right, will keep the edger moving in the correct direction and prevent it from gouging the floor.

An uncommon application for particleboard

by Howard Frank Itzkowitz

Floors are among the most highly visible parts of a building. Unfortunately, flooring materials and methods, like other building costs, constantly elude affordability. Those that are reasonably priced look as if they came straight from a discount store. Even worse, they're flimsy and don't wear well. We own a duplex (and live in the lower half), and when we needed to decide on flooring for a living and dining-room addition, we were committed to keeping an open mind. As a landlord and a home owner, I'm especially aware of the need to stay within my budget, and as an amateur carpenter, I have skills that are constantly stretched to their limits (and often beyond).

The floor had to be washable because part of the addition opens onto an exterior deck. We were concerned about food spills and wet shoes during the rainy season. It also had to be durable—maintenance is not one of my favorite pastimes. I wanted the floor to look respectable and appeal to renters. In a custom design, the owner's tastes are all that matters, but in a rental unit, I believe the floor should enhance anybody's furnishings.

The contenders? Though carpet was not ruled out, it wasn't a very good solution for us—we always seemed to prefer the rich colors and textures of the more expensive varieties. We liked the appearance of wood, and looked at everything from hardwood parquet to pine boards. Their prices were comparable to carpet. We ruled out ceramic, vinyl and rubber as not being appropriate for the room. What's more, they weren't cheap either. A knot began to grow in my stomach at the thought of dropping $2,000 and getting only a mediocre floor.

Then I remembered a material that I had used some time ago as a floor in a garage conversion. Particleboard is one of those things that we all love to hate. It's heavy, smells bad when cut, curls up like potato chips when wet and crushes when dropped on its corners; it doesn't like to hold nails or screws, and it's rough on sawblades. So why do we use so much of it for subfloors, countertops and cabinets? It's cheap, for one thing, but there are other advantages, too. Particleboard has a rich and interesting texture and color, not unlike cork, particularly when coated with a clear finish. It can be cut into small tiles or medium panels, or it can be left in its original 4x8 size. It's homogenous—no knots, checks, warping—provided it's kept dry. Waxing philosophical, thinking that new materials and techniques have to be explored if our homes are to be affordable, I decided to try particleboard again. One of the most difficult tasks is to use materials in ways that honestly and beautifully express their nature. This applies both to manmade materials and to natural ones.

I had designed the addition on a 4-ft. grid, and I felt that expressing the floor as a pattern would tie the room together visually and add richness. Just to ease my mind about moisture migrating up to the particleboard from the floor below, I stapled 15-lb. building paper to the subfloor. I splurged on some ¾-in. thick softwood

flooring, and ripped it into strips for the grid. These were laid out to chalklines and screwed to the plywood subfloor, and the screw holes were plugged.

The next part was the most difficult—working with the particleboard. I had the lumberyard cut it into 4x4 panels to make it possible for me to lift the stuff alone when I got it back home for the final trimming and setting. We gingerly stacked it on the roof of my Toyota, and even after dividing it into two loads it looked as if the car might turn into a pancake from the weight. I'd go to 2x4 panels next time. The work was hot and heavy with 4x4 panels, and they were

awkward to manage on a table saw. I trimmed and, with butt joints, fit all the panels as snugly as possible, then screwed them to the plywood, counterboring and plugging the screws in a pattern. Three coats of polyurethane topped it off, bringing out the rich amber color and texture of the material (photo above).

The result is durable and, I feel, makes a handsome and easily customized floor. Total costs for materials came to about $.60 per sq. ft.—a very affordable floor. □

Architect Howard Itzkowitz lives in San Rafael, Calif.

From *Fine Homebuilding* magazine (April 1987) 39:43

Laying a Plank Floor

Economical and good looking, single-layer flooring is also easy to put down

by Paul Hanke

Plank flooring is becoming more popular these days. In the northwestern U. S., many contractors use 2x6 T&G planks as a subfloor, and it's known as "car decking." Now that builders and home owners in many parts of the country are interested in post-and-beam construction, plank floors are chosen more frequently. In a floor that's framed with dimensioned 2x lumber, the floor joists are usually hidden above a drywall ceiling. But in timber-frame and post-and-beam construction, large wood structural members remain exposed in the finished house. Plank floors and widely spaced ceiling joists are a natural combination.

Plank-and-beam floors differ from conventional joisted plywood floors in several ways. The structural support in a plank-and-beam system is provided by relatively large beams that are spaced farther apart than joists. The second major difference is that the nominal 2x6 T&G planks (actual dimensions are 1½ in. thick by 5⅜ in. wide) form a single-layer floor that can be the finished ceiling when viewed from below. This is quite different from the multi-layer subfloor/underlayment/finish-floor system that's supported by closely spaced joists.

Plank floors can be less expensive than multi-layer flooring because you have fewer framing members to cut and assemble and because there's no need to install a drywall ceiling beneath a planked second floor.

There are some disadvantages to plank floors, however. Because the exposed plank floor will also be the finish ceiling of the room below, the quality of the material is important. Surface defects in the planks will be seen on both sides. Job-site damage is another potential problem. Once the floor is down, it has to be protected while the rest of the house is finished. If you're accustomed to tramping around on underlayment in the later stages of house construction, your work habits will need some adjustment.

Plank floors don't give you flexibility in locating wiring and plumbing because they don't provide hidden chases the way joisted floors do. Electrical and plumbing lines need to run in chases or wall cavities, or remain exposed in approved surface-mounted wiring channels. And in a finished house, many people feel that single-layer floors allow more noise transmission between floors. Also, some home owners (especially those with children) have noted that liquid spills upstairs quickly drip into the room below, despite the T&G feature.

In cases where you need to insulate beneath a plank floor, the insulation (fiberglass batts or rigid foam) can't be friction-fit between floor beams. Builders fasten twine, wire or 1x strapping across the bottom edges of floor beams to hold insulation. Other techniques for holding insulation include galvanized metal darts available from Insul-Mold (Civic Center Drive, Augusta, Maine 04330) and waxed cardboard trays from Insul-Tray (4985 North Cascade Place, Oak Arbor, Wash. 98277).

Design factors—The spacing between beams in a plank floor depends on the length of the planks and their deflection characteristics. Using 4-ft. centers for beam spacing is a safe, conservative approach if you're not sure of the deflection characteristics of your planks. Wider centers can be used for wood species and grades that are especially stiff, like Select Douglas fir (see the sidebar on p. 117).

The dimensions of the house have a lot to do with beam spacing beneath the planks. For example, a 36-ft. long house divides nicely into 4-ft.

or 6-ft. bays. A 38-ft. long house would divide up into awkward 6-ft. 4-in. centers for beams. With this spacing, you'd have a lot of unnecessary cutting, since 2x6 planks typically come in 12, 14 and 16-ft. lengths. If you have specially milled *end-matched planks* (planks milled with a tongue-and-groove on ends as well as edges), the ends of some planks do not have to be supported from below, so joist spacing is less important. But this "random lay-up" technique has its own limitations and isn't widely used. For one thing, end-matched planks are more expensive.

Concentrated loads can pose problems in plank-and-beam floors, and this is an important design consideration. If properly sized, a plank-and-beam floor will adequately support normal design loads. But items like bathtubs, refrigerators and waterbeds can cause excessive deflection. At the design stage, it's important to determine where concentrated loading will occur and either reduce spacing between beams or add blocking where necessary. This extra structural support can be attached to beams with standard steel hangers, or with timber-frame style joinery.

Many post-and-beam or timber-frame structures are designed without interior bearing walls. Nevertheless, non-bearing partition walls that run parallel with the planks should be framed up slightly differently. The bottom plate of a stud-wall partition should be built with the double 2x4s on edge, rather than face down (top drawings, next page). Use the same detail on interior bearing walls that run perpendicular to the planks. If a continuous bottom plate in such a partition wall is interrupted by a doorway, blocking should be installed beneath the floor between beams where the interruption occurs.

In addition to beam sizing and spacing, you

The structural part of a plank-and-beam floor consists of floor beams spaced on fairly wide centers. In this floor, 4x4 posts at midspan help to carry the load.

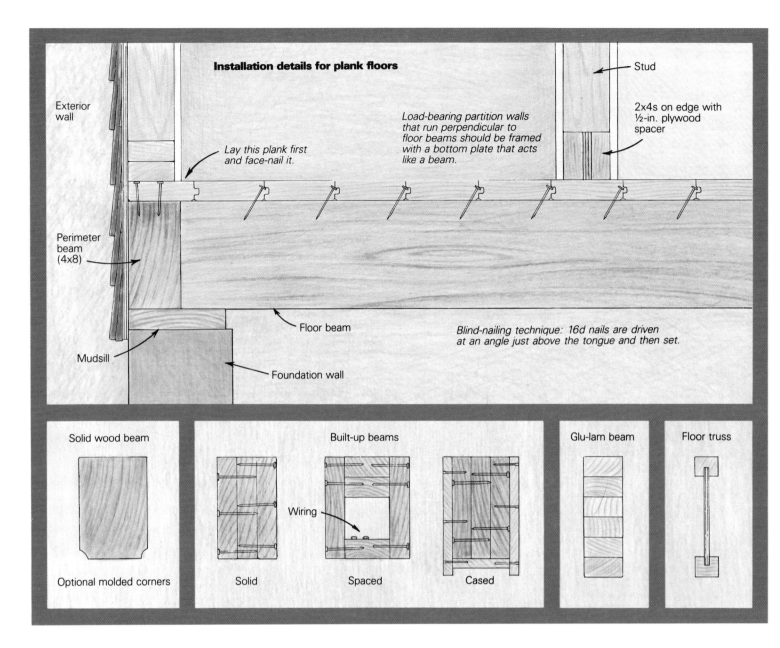

Installation details for plank floors

Stud

2x4s on edge with
½-in. plywood
spacer

Exterior
wall

Lay this plank first
and face-nail it.

Load-bearing partition walls
that run perpendicular to
floor beams should be framed
with a bottom plate that acts
like a beam.

Perimeter
beam
(4x8)

Floor beam

Blind-nailing technique: 16d nails are driven
at an angle just above the tongue and then set.

Mudsill

Foundation wall

Solid wood beam

Optional molded corners

Built-up beams

Wiring

Solid   Spaced   Cased

Glu-lam beam

Floor truss

have to decide on what type of beam to use: solid wood, beams built up from 2x lumber, cased beams, floor trusses or glue-laminated structural members (drawings, above). Cost, appearance, availability, load capacity and ease of handling are the factors to consider when making a choice. Solid wood beams are probably the most difficult to muscle into place, but they're also very attractive in the finished house. Built-up beams are a popular choice if they're spanning a crawl space or basement, where looks aren't important. Glue-laminated beams are expensive but very reliable in terms of span capabilities over long distances. Because they are engineered at the plant, you don't have to second-guess your calculations.

If a beam is to be exposed in the ceiling over living space, I like to give it a decorative bevel or bead along its bottom edges (drawing, above left). Apart from enhancing the ceiling's appearance, the rounder edge won't ignite in a fire as fast as a hard edge will.

Material and layout—The most common residential flooring planks are kiln-dried, T&G 2x6s. There are quite a few different wood species that are used to make T&G planks. Douglas fir is

common on the West Coast; northern white pine is popular in the Northeast; in the southern states, you'll find yellow pine. In terms of economy, you're usually better off using a local species. Regardless of species, there are two principal grades used in residential flooring: Select and Commercial.

Plank layout is important because it affects the overall strength and rigidity of the floor. The strongest layout (sometimes called a type D span) calls for the butt joints of the planks to be staggered from one course to the next, with each joint occurring over a beam. A plank that is long enough to span four beams (a 16-ft. plank over beams spaced 4 ft. o. c.) will be stiffer than the same species and grade of plank that spans only two or three beams.

In addition to designing for strength, you also want to minimize waste. As an example, let's suppose you want to build a 36-ft. by 24-ft. house with a plank-and-beam floor. You'll use 12-ft. long 2x6 T&G planks over floor beams spaced on 6-ft. centers so that planks will be stiffened by spanning three beams. In addition, joints in alternate courses should be staggered. In other words, the first course begins with a 12-ft. plank, the second course begins with a

6-ft. plank, and so on. This will minimize cutting and waste.

To figure how much material to order, multiply the square footage of your floor area by 1.17, then add 10% for trim and waste. When choosing material, make sure that it's dry, straight and of the quality you're looking for. On one job, we ended up returning about 25% of the planks because of unacceptable defects: split ends, wane, knotholes along the upper groove edge, and excessive bowing. It's best to see the material before you take delivery.

Nailing it down—Installing a plank floor is much like installing hardwood strip flooring (see pp. 106-111). Like hardwood flooring, planks have an up face and a down face. The down face has milled V-groove that gives an exposed plank ceiling a decorative look. Don't make the mistake of facing the V-grooves up. You may think the floor will look nicer, but these depressions will fill with dirt and debris almost immediately. I've seen more than one builder make this mistake, so watch out. It's a devil of a callback.

You start laying planks at the edge of the building, always running them perpendicular to the floor beams. If your band joist is straight

Drawings: Chuck Lockhart

(you can check it quickly with a string), simply align the groove edge of the planks with the outer face of the band joist. Face-nail through the top of this first plank course into the band joist so that the nails will be concealed beneath the first-floor wall plate.

All remaining courses except the last are blind-nailed, with the nail being driven into the corner where the plank edge meets the tongue. Each plank end should bear on a beam, and galvanized 16d box nails are the best fasteners to use for this job.

Nailing technique can take a little practice. As shown in the drawing at right, you should use two nails when a plank crosses a beam, and the nails should toe inward. This increases their withdrawal resistance and will help to prevent squeaky floors.

Before nailing the plank fast, it has to be snugged up tightly to the previously installed plank. Simply blind nailing and setting the nails will normally pull the plank up tight to its neighbor, just as it does with hardwood strip flooring. Slightly bowed planks (which can be fairly common) might need some extra coaxing. You can sometimes pound the bowed section tight by hammering against a beater board of short plank scrap.

If you've put down only a few courses of planks, bar clamps can be used to snug up a bowed section. Otherwise, you can drive the sharp end of a small prybar (I use a flat bar like the one shown in the drawing above right) into the top of the beam at a slight angle and very close to the edge of the bowed plank. Then pry the recalcitrant plank into place using the bar as a lever; hold it firmly until you drive in the nails.

Drive the nails as close to the edge of the plank as you can, being careful not to dent the exposed edge of the plank with your hammer. Then set the heads with a heavy punch or a drift pin. The heads have to be set far enough into the wood so that the groove in the next plank will seat tightly.

Plank ends and edges along perimeter walls can be face-nailed, but it's still good practice to toe these nails to discourage withdrawal. In theory, you could also face-nail planks that will fall under interior partition walls, but this is chancy. If you misjudge, the nail heads may be visible in the floor later.

Once the deck is down, it needs to be protected until the house is complete. Cardboard offers good temporary protection against dents and abrasion. Water can stain the planks, and even short-term exposure to direct sunlight can start greying the wood. Some surface discoloration can be removed by sanding, but you don't want to rely too heavily on sanding. Make sure that your crew and subcontractors understand that this "subfloor" is actually the finish floor.

Floor finish for T&G planks depends on taste and budget. Some people prefer to stain the wood a darker color before sealing and waxing or coating with polyurethane. Others like the natural appearance, which will darken slightly with age. □

Paul Hanke is an architectural designer and builder. He lives in Plainfield, Vt.

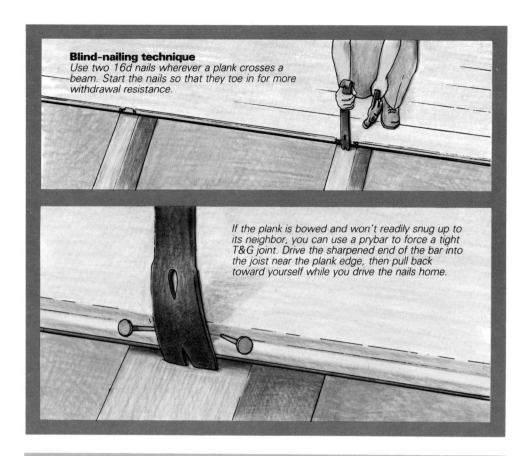

Blind-nailing technique
Use two 16d nails wherever a plank crosses a beam. Start the nails so that they toe in for more withdrawal resistance.

If the plank is bowed and won't readily snug up to its neighbor, you can use a prybar to force a tight T&G joint. Drive the sharpened end of the bar into the joist near the plank edge, then pull back toward yourself while you drive the nails home.

Species, grade and span: an elusive search

Installation simplicity and low cost are major attractions of plank flooring, but determining plank span (the spacing between floor beams) based on lumber species and grade can be frustrating. Though less stringent values have been used in the past, today most floors are designed to support a *dead load* of 10 pounds per square foot (psf) and a *live load* of 40 psf with a deflection limit of L/480, where L is the spacing between floor beams (in inches).

Flooring planks, which the lumber industry refers to as decking, have different span capabilities depending on the species and grade of the wood. Select and Commercial are the two most commonly used flooring grades. As shown in the sample chart below from the Western Wood Products Association (1500 Yeon Building, Portland, Ore. 97204), Select Douglas fir-larch will span 5½ ft. without deflecting over the L/480 limit with 50 psf loading. By contrast, Commercial lodgepole pine will span only 4¾ ft. to meet the same standard.

Search as I might, I couldn't find grade and species-specific plank-floor span charts for non-Western woods like eastern spruce, northern pine, eastern white pine, eastern hemlock and southern yellow pine. These are common decking woods in many parts of the country. Both the Southern Pine Inspection Bureau (4709 Scenic Highway, Pensacola, Fla. 32504) and the Northeastern Lumber Manufacturers Association (4 Fundy Rd., Falmouth, Maine 04105) have booklets that contain grading rules for regional lumber. But you won't find span charts for decking that relate species and grade to span capabilities. In fact, what you're given are engineering figures—specifically the *modulus of elasticity* and *extreme fiber stress in bending*—for different grades and species. These numbers have to be crunched in one or more formulas, depending on what type of span you'll have under your decking. Simple spans (sometimes referred to as *type A spans*), where planks bear on only two beams, aren't as strong as double *(type B)* or triple spans *(type C)*, where the planks bear on three or four beams, respectively. *Type D* spans are stronger still, since the butt joints in adjacent courses are staggered.

It's ironic that a floor system that's simple and inexpensive to install should be so quirky to design. This wouldn't be the case if the National Forest Products Association (1619 Massachusetts Ave. N.W., Washington, D. C. 20036) would update their existing plank-and-beam framing manual (Wood Construction Data #4), with some charts like those that the Western Wood Products Association has on species, grade and span for decking. Until such time, you might end up consulting with an engineer to figure out a good marriage between floor-beam spacing and plank species and grade.   —*Tim Snyder*

2x6 plank flooring: span capability based on species and grade*

Species and grade	Select Douglas fir-larch	Select hem-fir	Commercial hem-fir	Select lodgepole pine	Commercial lodgepole pine	Dense Sel. Southern pine
Maximum simple span	5 ft. 6 in.	5 ft. 2 in.	5 ft.	5 ft.	4 ft. 9 in.	5 ft. 5 in.

*Assuming a dead load of 10 psf, a live load of 40 psf and a deflection limit of L/480.

Ticksticking

by Sam Clark

Fitting a large panel—such as a section of plywood subfloor or countertop—into an irregular space isn't easy. It can be done by making a paper pattern, by laying out some sort of grid on which to plot points, or by cutting the piece oversize and then laboriously trimming away the excess. Often a better method is ticksticking, a nautical carpentry technique I learned from fellow builder Henry Stone, who discovered it in an old yachting magazine.

Ticksticking can be used to reproduce any flat shape quickly and accurately. It's good theater, too. You make some apparently nonsensical hieroglyphs on a scrap of plywood and a stick, and some equally arcane scratches on the stock to be cut. Then you saw out a shape without the intervention of ruler, bevel, level or mathematics. Your audience—which surely will have gathered by now, and which will have been making unkind comments at your expense—falls silent. Then there's applause, as the piece goes in the first time with no trimming, all 20 facets perfect.

The applause is entirely undeserved, however, because the method couldn't be simpler. Suppose you want to cut a countertop to fit against a wall that takes several jogs to form a niche at a window opening. No moldings will conceal the joints; the fit must be exact. Take a scrap of any thin sheet material; ¼-in. plywood is ideal. It's best if this scrap is at least one-third as big as the area to be measured. Secure it to the cabinet in the plane the counter will occupy. The scrap can be positioned anywhere on this plane, at any angle, but it is convenient to align one or more edges with the eventual location of the counter. In this case, the front edge of the scrap overhangs the cabinet 1 in. because the counter will eventually do the same.

Make a tickstick—just a thin stick or a piece of lath about 4 ft. long, with a point at one end. Lay the point on one of the critical lo-

Sam Clark is the author of The Motion-Minded Kitchen, *published by Houghton Mifflin (1983).*

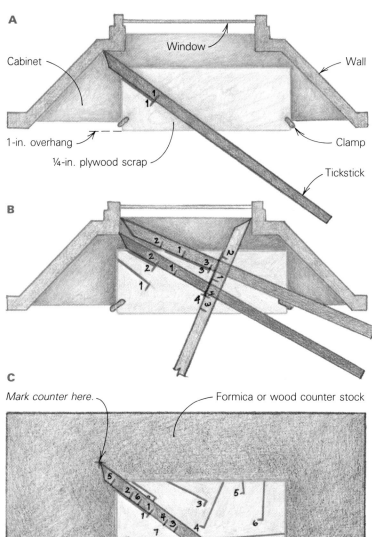

A

Cabinet
Window
Wall
1-in. overhang
¼-in. plywood scrap
Clamp
Tickstick

B

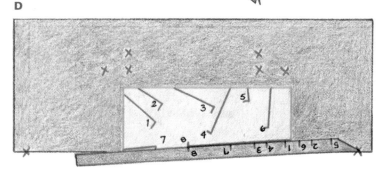

C

Mark counter here.
Formica or wood counter stock
Align flush.
Clamp
Plywood scrap

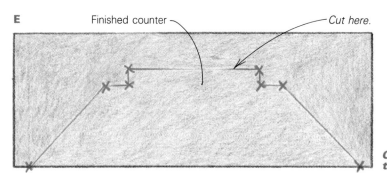

D

E

Finished counter
Cut here.

cations, say the left rear corner. Let the body of the tickstick fall anywhere on the scrap; it doesn't matter where. Hold the stick firmly, and with a sharp pencil draw a line on the scrap along the left edge of the stick. Without moving the stick, make a hash mark across the line you just drew and on the stick **(A)**, at the same point along the line. It doesn't matter where along the line you choose, as long as the two hash marks meet. Label both hash marks #1.

Mark the stick and scrap at first point.

Now reposition the tickstick, say with the point at the left corner of the window bay. Again mark along the left edge, make two more hash marks, and number them #2. In like fashion, mark and number all critical points along the perimeter **(B)**. For a curve, approximate by fixing many points along the arc.

Mark the other critical points.

You'll end up with a stick with numbered hash marks, and a scrap with lines crossed by numbered hash marks. Now remove the scrap from the cabinet, and clamp it on the countertop to be cut **(C)**. This can be done near the cabinet, or in your shop. To mark the tickstick, the scrap was centered in the space to be fitted with its front edge projecting 1 in. for the countertop overhang. To transfer these points onto countertop stock that is cut to rough length, just center the plywood right and left, with its front edge flush with the front edge of the stock. Now put the tickstick to the right of line #1, with hash mark #1 on the line touching hash mark #1 on the stick. Mark the counter right under the point of the stick. Do the same at line #2 and all the other lines **(D)**. Connect all these points, along with the one that marks the counter's length on its front edge **(E)**, and cut along the resulting line. Install the piece. Turn to the audience. Bow. □

Align the scrap and stick at #1; mark counter.

Align and mark other counter points.

Connect the dots.

From *Fine Homebuilding* magazine (April 1984) 20:49
Illustrations: Frances Ashforth

Floorboard styles. If you decide on custom flooring, you've got a lot of species, widths and edge treatments to choose from. Starting at the bottom of the stack: 10-in. antique yellow pine; 20-in. white pine—about as wide as you can find these days; 2½-in. quartersawn yellow pine next to 4-in. tongue-and-groove No. 1 Common red oak; 6-in. clear red oak; 9-in. sound (no holes) butt-edge red oak, and 4-in. No. 1 Common butternut.

Custom Wood Flooring

If you don't like the commercial offerings, here's how to find what you're after

by Paul Fuge

Wide boards and custom plank flooring are a breed apart from the strip and plank flooring available through normal retail channels. Commercially made plank flooring is a good value, but you get what the factory makes, which means you get what a lot of other people get. A well-informed consumer can specify custom plank or wide-board flooring to create something unique and interesting. To begin, you need to locate a willing processor and a suitable pile of lumber.

In standard lumber parlance, a plank is a roughsawn board 2 in. thick. This is hardly what the flooring-industry people mean. To them, a plank is T&G flooring 3 in. to 8 in. wide. Wide-board flooring is 8 in. wide or

wider, and is either butt-edged or shiplapped. Primary sources of custom-made plank and wide-board flooring are large-scale wholesalers with in-house millwork shops. They are capable of manufacturing hundreds of thousands of square feet of flooring a year. You can find these manufacturers listed in the Yellow Pages under Lumber—Wholesale. Even though your flooring order may seem big to you, it may be smaller than anything a big wholesaler wants to deal with. If this is the case, ask him who retails his products, or see if you can get him to recommend a manufacturer who will sell the amount of flooring you want. You can also try architectural millwork shops that deal directly with architects and

builders. They buy raw materials from wholesalers and mill it to specifications.

Also listed in the Yellow Pages are specialty lumber dealers who handle unusual building materials. These days, antique lumber dealers often carry new materials as a substitute for the increasingly rare high-quality old material they used to carry.

Materials—The most common species used for flooring are oak, maple, white pine and yellow pine (photo above). Many others are also suitable. Most hardwoods make excellent flooring if they are high grade—Select or FAS (lumber grading is explained on p. 121), and some, like soft maple, are often cheap. Readily

Floorboard edge treatments

V-groove

Back-relieved T&G

Wide-board shiplap

¼ in.

Lumber for flooring must be kiln-dried to a moisture content of 7% to 10%, because no matter how long boards may have been air dried, they will shrink once they're installed in a heated house. This is a small kiln operated by the author's specialty lumber company.

Unlike tongue-and-groove strip flooring, which is blind-nailed, most boards over 7 in. wide have square edges and are usually face-nailed or screwed to the subfloor. Here, a wide board is being sent through the planer. The first pass flattens the board before it's taken down to final thickness.

available kiln-dried Firsts and Seconds, No. 1 Face, Select and No. 1 Common grades of lumber are used full length, and many of the finished pieces are over 6 ft. long.

Most sawmills can cut logs for plank and wide-board flooring, but unless they have kilns, you are presented with the problem of properly drying the lumber before it is milled into floorboards. You want your flooring lumber kiln-dried to 7% to 10% moisture content (photo left). In most parts of the U.S., the only way to ensure that air-dried lumber won't shrink is not to heat your house. Don't be misled by those who say that if hardwood is air-dried for long enough it won't shrink indoors. Your best bet in this matter is testing a typical sample with a properly calibrated moisture meter.

Nearly all commercially available species can be sent through a molder-planer to surface and tongue-and-groove them. Your choice depends upon aesthetics and cost. Some hardwoods, like butternut and tulip, are soft enough to be dinged and gouged by the toenails of large dogs. Dryness and grade are the strongest determinants of satisfactory flooring performance. Grades of lumber lower than No. 1 Common tend to chip badly during manufacture. Common sense and a look at the lumber before manufacture will determine the suitability of any given species. The many books on wood and trees have descriptions of the hardness and typical uses of each wood.

If you're working with a sawyer or mill operator, talk to him about what you want, and see if he can make any suggestions. Press him a bit. He may know some beautiful local wood just right for your purposes. Don't forget about the logistics of drying: labor, technique, time and some sort of kiln, even if it is just a heated spare room or attic. Lumber dealers sometimes have odd lots of kiln-dried lumber tucked away that they'll sell at a reduced price. It's also not uncommon to find a species that isn't ordinarily used for flooring, but that might be wider or prettier than the usual choices. Mineral-streaked maple is a good example. Look at samples and let common sense be your guide.

Plainsawn vs. quartersawn—There are several ways a log can be sawn into boards. The most common are plainsawing and quartersawing. Because of the orientation of the growth rings in the board, there is less shrinkage across the grain with quartersawn boards than with plainsawn lumber, but to qualify the comparison, bear in mind that modern interior spaces rarely experience the wide swings in relative humidity and temperature common in previous centuries. Quartersawn lumber presents hard summerwood on edge as the wearing surface; however, only industrial floors really get the kind of severe wear that requires quartersawn lumber.

The choice between plainsawn and quartersawn wood is largely a matter of aesthetics and economics. Quartersawn flooring looks like a lot of straight lines, and it rarely comes wider than 5 in. Plainsawn flooring can be

Illustration: Lisa Long

quite wide, and it shows a greater range of grain patterns and figure. Only softwoods show a significant difference in wear between quartersawn and plainsawn boards in domestic applications, and only old-growth, narrow growth-ring quartersawn softwood is significantly harder on the quartersawn face.

Milling custom planks—Plank flooring begins as piles of lumber sorted by width. A straight-line ripsaw or a gang ripsaw transforms the crooked and kinked lumber into a pile of boards with straight parallel edges. The boards are then sent through a molder-planer, which simultaneously surfaces their tops and bottoms, and cuts the tongues and grooves on their edges. Some shops can make T&G to 11 in. wide. What emerges from an accurately set-up molder is a finished floorboard ready to lay.

Three patterns are typically available. Tongue-and-groove allows the use of either face, if the tongue and groove are centered along the edges. The second pattern, the V-groove, reduces the effects of machining variations, and the flooring can be prefinished. The price you pay is having V-grooves in your finish floor. Of course, you can turn the boards over and make it look like T&G. The third pattern is back-relieved to create air channels that distribute and dissipate moisture on the underside of the floor. Where the possibility of high moisture conditions exist—in a house over a dirt crawl space, say, or a beach house up on piers—back-relieving is important. The drawing at the top of the facing page shows the second two patterns.

Milling wide-board flooring—Flooring that's wider than 8 in. is a more adventuresome game. It's hard to find usable rough lumber this wide, because it requires considerably more skill to dry and mill it properly. Many millwork shops won't tongue-and-groove wide boards, but fortunately, the utility of the tongue-and-groove diminishes after 8 in., and most floors with boards this wide are face-nailed or screwed and plugged. Actually, the primary function of T&G matching is to allow the floor to be installed with no visible fasteners. Properly dried and milled lumber installed on a good subfloor with fasteners close enough to the edge (about 1 in.) will have little tendency to move. The idea that T&G will somehow help reduce shrinkage or the visibility of the crack is a myth.

Eliminating the tongue and groove eliminates the need for a molder and vastly increases the number of shops capable of producing a finished product. A planer and a straight-line ripsaw or gang ripsaw are the only tools required.

The procedure used to create wide-board flooring is to straight-line rip or gang-rip to standard widths, and to surface both sides, giving the greatest care to the best face. Wide boards are hard to flatten properly without cracking them along their centers. A skillful planer operator can do a decent job if he takes special care in putting the boards through the

planer. The best results are obtained with two passes with a two-sided planer or three with a single-sided planer.

In both cases the first skip pass flattens the board somewhat before it is planed to final thickness. If too heavy a cut is taken on the first pass, the powerful springs on the feed rollers of a wide planer will force the cupped board flat, often cracking it, before it passes through the cutterheads. As it emerges from the cutterheads, it springs back with its original cup, or, worse, lies flat with a broken back. Taking two passes diminishes the problem. It involves additional labor cost at the mill, but delivers a better finished product. Some millwork shops have special planers with softer feed rollers to help eliminate cracking. An alternate, and very labor-intensive, procedure is to run the boards over a wide jointer to flatten one face before sending them through the planer. Removing end checks and large knots near the ends before milling improves the quality of the finished product and eliminates arguments about the quality of the dimensioned lumber.

On wide-board flooring, the type of edge treatment will be determined by your needs and by the capabilities of the millwork shop. Butt-edge boards are most common. They can be manufactured on simple equipment and laid without trouble provided they are straight and flat; however, the cracks between the boards created by seasonal movement becomes more distinctly visible as the average width of the boards increases.

A wide-board floor naturally has far fewer cracks between boards to distribute seasonal movement than a strip floor. Shiplapping reduces the initial visibility of the cracks by making them half as deep. The shiplap (drawing, top of facing page), should not be more than ¼ in. wide. In properly dried lumber, the boards move no more than ¹⁄₁₆ in., and wider laps are weaker. From a practical point of view, any cracks will quickly and permanently fill with dust, so butt-edge soon looks just like shiplap flooring. Ceilings and walls, of course, are a different matter.

Flooring is sold by volume (board foot) or by area (square foot). Board foot refers to the nominal volume of the rough lumber that went into the manufacture of the finished product. One board foot is 144 cu. in. Thus a rough board 6 in. by 1 in. by 6 ft. long contains 3 bd. ft. This is a common but, in my opinion, misleading way to describe flooring. As far as I can see, its only advantage is to make the price per unit measure lower than if it were described by actual coverage in square feet. If the flooring is thicker than one inch, board-foot measure is an awkward way to calculate your flooring needs. Any seller of flooring can convert board-foot measure into the actual square footage that you need, and should do so at your request. Add 10% for installation loss, and you will know how much your materials cost. □

——————————

Paul Fuge has a wholesale lumber and millworks shop in Shelton, Conn.

Paul Fuge has a wholesale lumber and millworks shop in Shelton, Conn.

Lumber grading

No two boards are ever the same. Only a small percentage are suitable for flooring. Nearly all lumber is graded after it is cut into boards. The National Hardwood Grading Rules were developed to aid furniture manufacturers in determining the number of clear cuttings (wood without knots) available from each board. The grade of the rough lumber will give you an idea of what your custom plank flooring will look like.

Firsts and Seconds (FAS), considered the best grade, allows only a few knots or growth irregularities, and little discoloration. FAS makes the clearest, longest flooring, but it is also the most expensive and can be very bland. Select (Sel.) is like FAS, but can be as narrow as 4 in. and as short as 4 ft. In species with distinct sapwood, both FAS and Select may have visible sapwood on both faces. No. 1 Common (#1C) allows significantly more knots and growth irregularities. A combination grade called No. 1 Face (F1F) requires that one side of a board meet criteria for FAS while the other at least meets No. 1 Common.

The chart below shows typical finish lengths and widths, as distinct from the graded sizes above.

Grade	Surface quality	Typical lengths (ft.)	Typical widths (in.)
FAS	Mostly clear	6 to 11	4 to 6
F1F	More character than FAS, with fewer through knots than #1C	6 to 11	4 to 6
Sel.	Mostly clear, with sapwood	4 to 10	2 to 4
#1C	One or more knots per board	3 to 8	2 to 4

Occasionally, a pile of No. 1 or No. 2 Common lumber will be perfect for flooring. Aggregations of tiny knots called cat's faces can degrade white oak boards all the way to No. 2 Common even though the board looks nearly clear. Beautiful mineral staining in maple can degrade clear boards to Common. A look at a specific pile of rough lumber is the best way to determine which grade is suitable. Examine any grade marked No. 1 Common or below board by board.

You can always step outside of the NHGR and deal directly with a small sawmill. You may find a sawyer who has set aside wide boards on sticks awaiting your arrival. Just be aware that after years on sticks wood gets no drier. The reassurance that "the wood has been drying for two (or ten or forty) years," will do little to prevent it from shrinking further once it is indoors. Only continued drying on sticks in a heated space or kiln will finish the job. —P. F.

Swedish Floor Finish

Why a hardwood flooring specialist likes this European finish, and how to apply it

by Don Bollinger

Although there are many brands of floor finishes on the paint-shop shelves, they can all be divided into two basic types—surface finishes and penetrating finishes. A surface finish, such as a urethane or varnish, places a thin, durable skin on top of the wood to protect it from moisture and abrasion. In contrast, penetrating finishes are absorbed by the wood until numerous coats create a surface buildup. They create lustrous surfaces that show off the color and figure of the wood without imposing a slick-looking layer of plastic over it. Penetrating finishes rely on the hardness of the wood for durability, and therefore they won't wear well on softwoods.

Usually my clients request a floor finish surface that combines the best qualities of both types of finish. While they are more closely related to the surface finishes, the floor coatings called Swedish finishes combine the easy maintenance and durability of a surface finish with the appearance of a penetrating finish. They don't hide or fill in wood grain as much as most varnishes, waxes and urethanes unless excessive amounts are applied. They wear extremely well, they never need waxing and they resist moisture and nearly all household chemicals and stains. They don't yellow excessively with age, and unlike most varnishes and urethanes, will not peel or fracture from dimensional changes in wood due to seasonal moisture variations. This resiliency makes a Swedish finish an especially good choice for softwood floors, such as pine or fir, which expand and contract more than hardwoods. The porosity of softwoods also allows a Swedish finish to penetrate deeply, strengthening the wood fibers that absorb it. What's more, most folks find a Swedish finish surface to be less slippery than most other surface treatments.

Different types of Swedish finishes—Every manufacturer of Swedish finishes uses a different formula to concoct its products, and each formula is a closely guarded secret. Manufacturers are willing to admit, however, that the principal ingredients are alkyd urea resins suspended in a solvent. For 40 years, the primary solvent has been alcohol. But now the manufacturers are heading toward water-base finishes, which eliminate the cloying stench of formaldehyde that offgases from solvent-base finishes. While I expect to be using water-base finishes sometime in the future, I'm still not convinced that they are as desirable as their solvent-base ancestors. This article is about solvent-base finishes.

Swedish finishes are designed to give a com-

A wide brush made especially for applying Swedish finishes makes coating a floor go quickly. A Rubbermaid dishpan makes a good reservoir for the finish during application, and wrapping it with a trashcan liner speeds cleanup time. Before finishing, this white oak floor was tinted with a vegetable-dye stain suspended in a tung-oil base.

From *Fine Homebuilding* magazine (December 1986) 36:42-45

pleted surface in two applications, although you can apply more layers to increase the thickness, and hence longevity. We usually apply three or four coats to floors in commercial buildings. Various brands are available from suppliers in the U. S. (see the list of addresses on the next page). We use either Futura or a combination of Bacca and Glitsa. Futura and Bacca have to be mixed with a catalyst before application; Glitsa can be used right out of the can. For more on the materials and equipment we use for applying Swedish finishes, see the sidebar at right.

The Bacca/Glitsa combination is an anomaly among other Swedish finishes. All the others are applied in several coats—the last is the same as the first. Glitsa, on the other hand, is a finish coat used in conjunction with Bacca, which acts as its undercoat. Bacca can also be used as a finish, but it comes only in a high-gloss version. Almost all of my customers want a satin finish on their floors, and Glitsa comes in semi-gloss. If I need multiple layers, I apply more Bacca. Applying Glitsa directly over Glitsa within a 90-day period can lead to any number of complications, including complete peeling of the entire surface coating.

Most professionals I know like to use the high-gloss version of their favorite brand of Swedish finish as the bottom coat or coats, followed by a finish application of the same brand in semi-gloss or matte finish. The particles used in the finishes to give them a semi-gloss or matte appearance reduce the durability of the film. The fewer particles overall in the various coatings, the longer the life expectancy of the floor.

While I prefer the look of Futura, I routinely use the Bacca/Glitsa combination because it wears better. I have seen at least a half-dozen different floor-finish abrasion tests conducted by independent labs, and invariably some combination of products with Glitsa as the top coating wins out over all the others.

Sanding, filling and cleaning—If you have a newly installed floor, it should be sanded with progressively finer grades of sandpaper until the entire surface has been sanded with 80 to 120-grit paper ("Floor Sanding," p. 112). Most waxes, urethanes and varnishes will hide the minor imperfections, such as sanding marks and small voids, found in nearly all wood flooring. The Swedish finishes will not—it highlights them. A poorly sanded floor will definitely show its flaws with a Swedish finish on top.

When we sand new floors, midway through the process we fill nail holes and minor voids with a mixture of lacquer-base wood putty and extremely fine sawdust that matches the species of wood. But this kind of cosmetic touchup should be kept to a minimum. The time to eliminate cracks and holes in your wood floor is when you are selecting your flooring material and installing it. A higher grade of stock and a few more nails will go a lot farther toward keeping your floor looking young than the use of any wood putty.

If you're going to use Swedish finish on an old floor (especially if the floorboards show signs of movement), I don't think you should use any putty at all. Instead, let the finish flow into the

Tools, materials and gear

Swedish finish—Get enough finish to coat your floor at least twice. Most Swedish finishes will coat 200 to 350 sq. ft. of flooring to a gallon of liquid, depending on the species of wood and whether or not the grain has been sealed before coating.

Brushes—I prefer to use the extra-wide floor-finishing brushes developed by the Swedish finish manufacturers (photo facing page). Before using one, I condition it by soaking it in lacquer thinner or denatured alcohol for several days. During this time I dry it out and brush it occasionally to remove any loose bristles. Brushes should be thoroughly cleaned and dried immediately before and after each usage.

Solvents—Use denatured alcohol for Futura and lacquer thinner for Bacca/Glitsa.

Respirators—Tests conducted by a representative of the Washington State Department of Labor and Industries show that formaldehyde and other organic-vapor exposures exceed U. S. Government limits for short periods of time during the application of Swedish finishes. For this reason, it is imperative that you wear a respirator when using Swedish finish. It should have a cartridge/filter specifically designed for use around formaldehyde. MSA (Mine Safety Appliance, Dept. L, 600 Penn Center Blvd., Pittsburgh, Pa. 15235) makes such a cartridge.

Rags—You'll need a few rags for cleaning brushes and equipment. Leave one or two dampened with a little solvent at the front door or near the area to be coated. They can come in handy while you are applying the finish. To avoid problems with spontaneous combustion, we store rags soaked with solvent or stain in a 55-gal. oil drum partially filled with water.

Floor buffer—This is not absolutely necessary, but for large floors it can really

speed the work. Be sure to get a pad pusher and a buffing pad.

Sandpaper—A good floor finisher always carries a piece of sandpaper while finishing for final touchups or an occasional "oops."

Screenback—A screenback is an abrasive disc that resembles insect screen. One 120 or 180-grit screenback to fit the buffer you are using should do the job in most cases. Normally one side of the screen is good for about 500 sq. ft. of floor. Screenbacks are very expensive, so don't buy more than the job requires. If you don't want to use a buffer you can use burnished sandpaper in the last grit used to sand the floor. To burnish it, rub two pieces together to remove the hard edges. Hand-sanding takes more time and elbow grease, but the results are good and it saves money.

Sweat band—Lots of folks scoff at me for it, but I know I sweat when I'm working, and a $.99 sweat band keeps me from dripping sweat in the finish. Sweat droplets will cause a white spot in the finish that usually turns into a bubble when the white fades away. If you do happen to find such a spot in the finish, quickly and lightly hand-sand it out. There is usually no need to attempt to clean up the residue from this light sanding. Simply coat again right over it.

Overalls—I like to wear painter's overalls when finishing floors. They help keep my other work clothes free of Swedish finish.

Knee pads—Maybe it's because I'm getting old, but a long time on my knees without knee pads is out of the question. I can always do a better job if I'm comfortable.

Work shoes—Wear soft, white-soled shoes that will not track in dirt or scratch the floor. Remember that any scratches in the freshly sanded or stained surface will be there until the next time you refinish.   —D. B.

cracks between the boards. When the boards move, the finish will crack beneath the surface, out of sight. Obviously the floor won't look as good as a new one, but years down the road it won't have putty popping out of the cracks and taking the finish with it.

Swedish finishes are incompatible with other finishes. Consequently, if you're refinishing an old floor with another type of finish you'll have to take it down to bare wood. Stains from pet urine, potted-plant leaks and other moisture damage may not sand out completely. These do not normally interfere with adhesion, although their looks in the final product could be bothersome. If you are sure the top layer of old finish on your floor is a Swedish and you are certain that no waxes or oils have been used on it over the years, you can "top coat" your floor with one or two layers of new Swedish finish after a thorough cleaning.

Once the floor is sanded, you will need to sweep or vacuum it. I prefer vacuuming. First I use a crevice tool to clean the cracks, all the

junctions between walls and floors, under cabinets and appliances, and the tops of base moldings and window sills. If you haven't covered the heating ducts, be sure to vacuum them out. Should the heat kick on while you are finishing, you may find a whole lot of particles mired in the finish around the duct. I finish vacuuming the rest of the floor with a hard-surface nozzle with a brush.

When I'm done vacuuming, I like to "tack" the floor to remove any remaining dust. Many Swedish-finish manufacturers caution against this because you might introduce an oil residue that can interfere with the bonding of the finish to the wood. To prevent this problem I dampen the tack rag with the solvent recommended for the finish I'm using—denatured alcohol for Futura, lacquer thinner for Bacca and Glitsa. Never tack with paint thinner. Always be certain that all traces of the solvent have evaporated before you apply any finish.

As we begin the final stages of sanding and cleanup, we start to regulate the temperature of

The first coat of finish will raise the grain. A floor buffer with a screenback disc (left) is a good tool to knock down the high spots. The screenback mounts against a pad that is used in place of the buffing wheel. Above is a stained and finished, white oak floor.

Brands and suppliers

Bacca/Glitsa: Glitsa American Inc., 327 S. Kenyon, Seattle, Wash. 98108.

Futura: Bowen Jiffy, P.O. Box 216, Mamaroneck, N. Y. 10543.

Nordic: UPP Inc., 2442 1st Ave. S., Seattle, Wash. 98134.

Skandian: Minwax Co. Inc., 102 Chestnut Ridge Plaza, Montvale, N. J. 07654.

the interiors to be finished. They should be at least 65°F. If you are going to use a stain on the floor, now is the time to apply it.

Staining with Swedishes—Swedish finishes were originally developed to be used directly on raw wood. Applied this way, the first coat penetrates much as any sealer would. When subsequent coats of Swedish are applied, they partially "melt" the surface of the preceding one. When all the coats are dry and fully cured, the floor will have a continuous bond from its surface down into the wood itself.

If you stain your floor, you are sealing the grain in the process. As a result, the first coat of finish must rely on variations in the texture of the wood and a slight chemical bond for a footing. This can cause a loss of continuity at the surface that won't show up until the floor's surface is traumatized. For instance, a dropped piece of firewood will dent the floor. With an unstained floor the finish and the wood will dent equally, and usually stay bonded. In the case of the stained floor, chances are there will be a slight separation between the finish and the

stained surface. This blister can break, and the finish will begin to peel away from the wood—especially with pigmented stains.

Although the original formulas for Swedish finishes have been altered to make them compatible with stains, you've got to be careful to avoid adhesion problems. Your best bet is to use a stain recommended by the manufacturer of the floor finish. Lacking this sanction, you will probably be safe in sticking to a vegetable-dye stain that is linseed-oil or tung-oil based. But even among these stains, some of the chemicals used to create a specific color can cause problems. If you are in doubt about a stain, make a test patch. When the finish dries, apply a piece of duct tape to it. If the tape pulls off the finish when you remove it, the stain fails the test. Stay away from petroleum-base stains, and avoid heavy dyes or pigmented stains. Stain/finish compatibility has become a particularly troublesome problem with the increased popularity of grey and white stains. These are heavily pigmented with titanium dioxide, which isn't compatible with Swedish finishes.

A big key to the success of using any stain under a Swedish finish is drying time. The more

porous the wood, the deeper the stain will penetrate and the longer it will take to dry. I find that most light stains will dry sufficiently overnight to take a finish the next day if they are thoroughly wiped down shortly after application and if there is sufficient heat and air circulation. Stains with pigment can take a week or longer to dry. If I'm in doubt about the dryness of a stained floor, I'll scrub it in a few suspect places with a clean, dry white towel. If there is any residue, it's still too wet to coat.

Final preparation—Swedish finish should be at or near the floor temperature when it's used, so make sure you bring the finish into the room at least 24 hours before you apply it.

Just before you brush on the finish, make sure all local pets are accounted for—especially cats. Nothing is more in their nature than to hide somewhere in the house, only to reappear on the wet floor with a puzzled look on their face and goo on their paws.

During warm weather, a Swedish finish will dry faster than on a cold, wet day. If this is the case on your job, ask your supplier for an extending agent. Added to the finish before application, the extender will retard its drying time, allowing you to avoid lap lines and brush marks.

Most good finishers work when there's no sharp, direct sunlight on large portions of the floor because it causes the finish to cure faster than they can work it. I sometimes drape a cloth tarp across windows to prevent this. If this isn't possible, I cover the floor with the tarp until I'm ready to coat it.

The first coat—Once the floor has been prepared for finishing, extinguish any nearby open flames. To keep the dust down and to retard the finish's drying time, close all windows and doors. Any drafts will cause the finish to set too rapidly, creating an irregular surface. Don your respirator and mix the catalyst into the finish you'll need for the first coat. Make sure that you mix enough—it's better to waste a little than to have to go back and mix more to complete the job.

Mull over the configuration of the rooms to be coated. You'll want to work away from the light as much as possible so that you can constantly see and critique your work. Always paint with the grain of the wood (photo p. 122). I prefer coating with a 12-in. to 14-in. brush, but many professionals swear by lamb's-wool pads or even flexible trowels. Paint a long row a few boards wide. When you get to a wall, go back and start in on the next run. Do only as much area as you can cover and still wet-lap each finish line by 6 in. or so. Do not allow the finish line to dry or you will end up with a lap line. Never flip the brush at the end of a stroke. Long smooth strokes ending in gentle feathering into the finished part of the floor are usually best.

As you begin a new line of finish, take sight of it from several different light angles, checking for lap lines, brush marks, puddles, drips and bits of debris. Remove bits with your fingers and stash them in your shirt pocket. Then lightly go over the spot where you removed the debris. Check again for "holidays"—porous woods drink up the liquid and leave gaps in the finish where you would swear you coated heavily.

Keep the brush or applicator over the bucket when you aren't using it. Drip marks and puddles will form under unwatched brushes and lamb's-wool applicators in the wink of an eye. And don't allow any water to get on the floor. If it does, immediately hand-sand it out with one or two light strokes. You'll be seeing it for a long time if you don't.

If two or more of you are finishing the floor at the same time, you should start in different areas and work toward each other. Be sure both of you are using the same mix—if they are even slightly different you will see it where they meet. Before you move out a doorway, make sure you've coated behind the door.

When the floor is completely coated, lock the doors or entrances into the wet areas and post signs if there is any likelihood of foot traffic. Fortunately the smell of the finish is usually enough to keep most people away until it is dry. In some cases you should tape shut the mail slot. I remember one job where the entire front page of the weekend newspaper was glued to the floor just inside the front door. I still don't know how the paper boy squeezed the edition through the slot, but he did.

When you finish applying the first coat, empty the unused finish into an airtight container and save it for later—it will retain its workability for several days. Don't pour it back into your uncatalyzed finish, and don't mix it with future batches. If you want to toss the unused finish, dispose of it in an airtight container or set it aside until it hardens. We have all our old finish and solvents picked up by a toxic-waste disposer.

Clean your brush or applicator at least twice in the emptied bucket with the appropriate solvent. This is a good time to clean the paint bucket as well. Several of my crew like to use a clean garbage bag as a liner for their buckets. Using a liner eliminates having to clean the buckets and reduces the amount of solvent needed for cleanup.

The second or final coat—Nearly all Swedish finishes will dry to the touch in one to four hours, depending on the temperature, humidity and air circulation in the room. We often wait overnight to apply the second coat. If the finish feels dry to the touch but you are uncertain about its readiness for recoating, try this simple test: place your hands firmly on the surface and press for 5 to 10 seconds. If your hands stick even slightly, you should wait another hour or so.

When coming back into a room to prepare it for the last coat, I put on my respirator and open all the doors and windows I can reach. Good cross-ventilation should make the area bearable without a respirator within 30 minutes to an hour. I check the floor thoroughly for lap marks, brush marks, puddles and the like. As I spot them, I remove them with a burnished piece of sandpaper the same grit as the final sanding grit or with a razor-sharp scraper. I then hand-sand all the corners and recesses. This is also my last chance to set any visible nails and fill their holes.

The first layer of Swedish finish will raise the grain of the wood slightly. A light sanding by hand or with a screenback (an abrasive disc that looks like a piece of insect screen) and buffer will smooth the surface nicely, preparing it for the final layer of finish.

I then prepare my floor buffer with a burnished 180-grit screenback, backed by a white or red 3M buffing pad mounted on a pad pusher. I like to start in a closet or area of the room not struck by a good deal of light because a relatively new screenback will sometimes leave circular marks on the floor that won't be totally covered by the final coating of finish.

When "taking the tops off the floor" in preparing it for the final coating, the idea is to develop a white powdery dust but not sand through the finish (photo facing page, left). Usually once or twice over a given area will do the trick. When you have finished buffing and hand-sanding the floor, repeat the vacuuming sequence. If I am applying three or more coats, I sand before each new coat is applied.

The final coat is applied much the same as the first coat, only there can be no mistakes with lap lines, puddles and drips. If you are using Glitsa you won't be adding a catalyst, and it should be applied in a thicker layer than the Bacca. In Europe many finishers use a highly flexible metal trowel for applying the finish.

The final coat will take longer to dry than the first coat. If you are using Glitsa and find a blemish, you have about 24 hours from the time it was applied to fix it. After that the repair won't bond properly. Other finishes can be repaired at your leisure, but are best done within 72 hours.

When a third coat of finish is desirable, we do a second coat of Bacca and let it dry overnight before screening and applying a layer of Glitsa. An even better three-coat system is to do the second and third coats with Glitsa, but to do so the manufacturer recommends a 90-day curing time between second and third coats. Most clients don't want to put up with the hassles involved in this method.

Recoating a Swedish finish—When a floor covered with Swedish finish begins to look dull and becomes difficult to clean, it's time to recoat. Unless the finish has worn completely through, you don't have to sand the floor to the wood. Instead, scrub the entire floor with lacquer thinner or denatured alcohol to remove ground-in dirt. Have several clean towels handy to mop up the dirty solution before it dries. Be sure to wear a mask with an organic-vapor filter. Next hand-sand the dry floor with 80 to 120-grit sandpaper or screen it with a buffer, followed by a mopping with a solution of two cups of vinegar to a bucket of water. Allow the floor to dry thoroughly, then vacuum. Final-coat it with Swedish finish as though it were a new floor.

Care of a freshly coated floor—A new Swedish finish is like a new slab of concrete—it gets substantially stronger and harder with age. Still, it's usually safe to walk on one after three to four hours drying time. I recommend stocking feet for the first 12 hours and soft-soled shoes for the first three days. After that period it is normally safe to set furniture on the floor, but you should leave throw rugs off for several weeks.

When you are ready to move furniture back onto the floor, be extra careful with heavy objects that have small feet or wheels, such as pianos, refrigerators, dining tables and couches. I use a rug or old carpet remnant turned upside-down to slide these bulky items into the room. Once they are in place, set them on coasters.

In heavy traffic corridors, a temporary runner made of two layers of corrugated cardboard will protect a new floor from a multitude of hazards. Don't use runners made of plastic or any other material that will retain air or moisture. They will retard the curing process.

Once the floor has cured, use a vacuum or an untreated mop for routine cleaning. For heavy-duty cleaning, mop it with a solution of two cups of vinegar to three gallons of water. For kitchen or dining-room areas, I've found that adding a small amount of mild detergent like Ivory Liquid to this solution will help degrease the surface. Frequent degreasings are not usually necessary except in restaurants. □

Don Bollinger is the owner of Oak Floors of Greenbank in Seattle, Wash.

Installing Baseboard

There's a tad more to it than coping the joints

by Bob Syvanen

Baseboard installation is often done badly. Why? Probably because it comes at the end of the job, after the crowns and casings, and carpenters are anxious to wind things up so they can get on to new projects. Or it may be because it's uncomfortable work done on hands and knees, with a lot of getting up and kneeling down. But maybe it's just because a lot of carpenters don't know how to do it right.

Shapes and styles—Baseboards are used to cover any gaps that occur at the juncture of walls and floors, and they also protect the lower wall from dings and scrapes. Visually they give weight, definition and presence to the wall, working with the crown molding and corners to "frame the wall." Baseboards are usually made of the same wood that's used for trim elsewhere in the house, and they can be either hardwood or softwood. The central part of the back face of baseboard stock is partially relieved, or plowed away, like casings; this helps it lie better against the wall. Baseboards, however, are usually thinner than casing stock. This is because casing is frequently made with a rounded outside edge, and a somewhat thinner baseboard can be butted against this edge without its looking awkward.

Standard baseboard comes in a variety of shapes and sizes, and custom shapes can be made in the shop with a table saw, router or shaper. Another way of getting a unique baseboard profile is to assemble it from combinations of standard moldings, as shown in the drawing below. There's really no end to the shapes that can be achieved when two or three-piece combination baseboards are used.

Coping the joint—Many of the techniques for cutting, fitting, nailing and finishing baseboard are similar to those required for casing. But the miter joint used so frequently for casing tends to open up on the inside corners of baseboards. A much better baseboard joint for inside corners is the coped joint.

A coped joint requires a different cut on each of the two boards to be joined. It involves some miter cutting, so a backsaw and miter box or a power miter box are required. Though I've cut miters by hand for years, I like the power miter box because it's fast, but doesn't sacrifice quality. You'll also need a coping saw. This small tool with a spring-steel frame looks like a C-clamp with a wood handle, and has a slender, fine-tooth blade stretched across the mouth of the C.

Cutting the first board in a coped joint is easy—just cut it square so it fits tight into the corner of the wall. The second board is coped. To begin a coped joint, miter the board vertically, as if you were going to make an inside mitered corner. When you're done, look closely at the front edge of the cut—it will reveal the baseboard's profile, and will serve as a guideline for making the second cut on the board. I rub the edge of the cut with the edge of a pencil lead to make it more visible.

To complete the cope, support the board, front face up, so that the end to be cut hangs just beyond some solid support—a workbench, sawhorse or cricket (a cricket is a portable step turned mini-workbench). Then, with the saw-blade nearly perpendicular to the bottom edge of the board, cut along the pencil line, following whatever curve is indicated (drawing, facing page, left). While cutting, incline the saw slightly to put an angle on the cut. The angle should slope away from the front surface of the board, and will help the lead edge to make good contact with the square-cut board when the two are brought together. If the baseboard has a flat top edge (like the one in the drawing), this edge should be square cut—an angle would show as a gap. If you've made the cut correctly, the end of the coped baseboard will slip right over the square-cut end of the one you installed earlier. This technique will work on just about any baseboard, and can also be used to fit ceiling molding. It sometimes requires a little adjustment with a sharp chisel or utility knife.

Installing standard baseboard—If there's a simple choice between a long, unbroken wall and a short one, I start the installation with the long one. It's easier to get a good fit with a long piece of baseboard than with a short one, and you'll see why in a moment. I also try to minimize any possibility that people will see a poorly fitted corner joint (if one happens to slip into the job). To do this, I like to install the first length of baseboard on the side of a room that's opposite the door. The baseboard on the adjacent wall will conceal the imperfect joint so that it won't be visible when someone first enters.

Let's assume that you've chosen to start on an unbroken wall that can be fitted with a single length of baseboard. Begin by measuring the wall, making sure to take your measurements at the floor level. Walls aren't usually in perfect plumb, so the measurement will vary depending

When baseboard must be fit to door casing, a measuring block (photo left) makes it easy to mark the baseboard to length. Built from scrap wood to fit the particular baseboard being installed, it is placed against the outside edge of the casing and over the baseboard, and a cut line is then scribed on the baseboard.

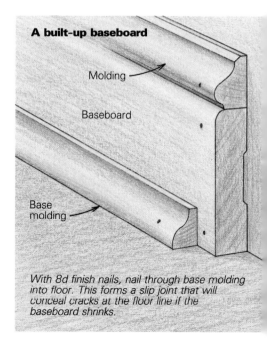

A built-up baseboard

Molding

Baseboard

Base molding

With 8d finish nails, nail through base molding into floor. This forms a slip joint that will conceal cracks at the floor line if the baseboard shrinks.

From *Fine Homebuilding* magazine (August 1986) 34:40-41
Photo: Pat Syvanen

on where you take it. Select a straight length of baseboard and clean up one end by cutting it square. Use the wall measure you took to find the other end and cut it square also. The cut should allow a snug fit, particularly at its top edge where it will be visible. Test the fit on the wall, trim off a tad if necessary and nail the piece in place. If you have to splice lengths of baseboard on long walls, a vertical scarf joint (overlapping 45° miters) is the best to use.

Nail through the baseboard and into the studs, using 8d finish nails top and bottom. Use as many nails as necessary to get the baseboard to pull tight to the wall. Studs can be located before setting the base in place by probing with a nail driven through the finished wall (the baseboard will cover the holes). Mark the locations above where the top edge of the baseboard will be with a light pencil mark so they won't be obscured when you set the baseboard in place.

With this done, you can begin work on the adjacent length of baseboard. This one gets a coped cut at one end and a square cut at the other (this same combination of cuts is used when a baseboard has to be fit between a door casing and another baseboard). First, cope the end that will butt against the board you just nailed in place, and check it for a good fit. Then measure to locate the square cut at the other end. The fit should be snug, but not so tight that it cracks the plaster or drywall when you nail it in place.

To get just the right amount of snugness on long boards, cut them a tad long so that the middle of the board will be bowed out from the wall when you put it in place. The amount of bow depends on the length of the baseboard you're installing, but it's usually about a finger's width, measured midway between the ends. But you shouldn't have to force it—a gentle push and the baseboard should snap into place as it nears the wall. Take particular care when you're fitting baseboard to the casing of a door. As you spring the baseboard into place you don't want to push the casing out of position. If the fit looks good at both ends, nail it in place. Keep work-

ing your way around the room until all the baseboard is in place. There may be times when you have to fit a baseboard between baseboards on opposite walls. If you do, just cope both ends and snap it into place.

For fitting baseboard to door casing, I use an L-shaped block (photo facing page) to help me mark an accurate cutline. By holding the top of the block against the edge of the casing and marking down the leg onto the baseboard, I get a precise measurement. With one end coped, slip the baseboard into place, allowing it to run slightly long past the casing on the other end. Place the block against the outside edge of the casing and over the baseboard, and mark a vertical line for the square cut. A cut made a tad outside this line will give you just enough extra length to spring the board into place.

Other joints—Coping doesn't work on all baseboard profiles. Simple rectangular baseboard, the kind with a rounded upper corner, just doesn't look good with coped joints because the abrupt curve makes for a very fragile overlap on the adjacent baseboard. Instead, I use a combination butt and miter joint for inside corners (drawing, below right).

The trick here is to make a mitered lap joint at the top edge of butting baseboards. Begin with a square cut at the end of one baseboard. Miter the top edge with a finish saw, cutting to the point where the rounded corner ends, and follow this with a cut at 90° to the first. This will release a triangular piece of wood. At the end of the second baseboard, make a miter cut that corresponds to the first, again just to the bottom of the rounded edge. Then turn the baseboard over. Working from the bottom edge, make a square cut that's angled just a tad away from the front of the workpiece. This cut should be stopped at a line scribed off the baseboard already in place, because there isn't much room for adjusting this joint. This is a much better-looking joint than a butt joint or a coped joint, though a butt joint can be okay when molding is used on top of it.

Outside corners—Outside corners on baseboard are always mitered. Since walls rarely make a perfect corner, I always make trial cuts to find the right angle for the miter. For a good fit, the angle cut on both pieces should be the same. I make my first cut long and gradually trim it to perfection using the miter box or a block plane. (With a miter box, shim cardboard or even plane shavings between the miter-box fence and the back of the baseboard; make fine adjustments to the saw angle by moving the shim away from or closer to the saw blade.) Outside corners are cross-nailed to lock the joint in place.

More tips on nailing—Occasionally you will be faced with a situation where the baseboard has to be pulled in against the wall, but there isn't any stud to nail into behind it. As an alternative, drive a 16d finish nail through the baseboard, angling it down and out so that it catches the 2x4 bottom plate in the wall. When the nail is set, the baseboard will be pulled snug against the wall.

In similar situations at inside corners and door openings, 16d finish nails again come to the rescue. Just angle the nail until you hit something solid. Keep it a few inches from the end of the baseboard and predrill the hole before you nail to prevent splitting. If you are framing up a new house, it's a good idea to install short lengths of 2x4 baseboard blocking (offcuts or scrap pieces) at all inside corners and at each side of door openings.

If you're installing a two or three-piece baseboard, the lower molding should be nailed to the floor, not to the baseboard. If it isn't, any shrinkage of the baseboard will pull the molding away from the floor and expose an unsightly crack. When it's nailed to the floor, molding serves as a slip joint, concealing any shrinkage cracks. Paint the baseboard before installing this molding so an unpainted strip won't show up after the baseboard has shrunk. □

Bob Syvanen, of Brewster, Mass., is a consulting editor of Fine Homebuilding *magazine.*

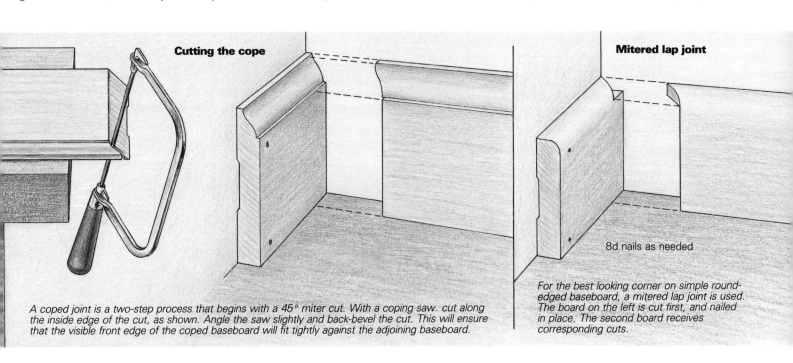

Cutting the cope

A coped joint is a two-step process that begins with a 45° miter cut. With a coping saw. cut along the inside edge of the cut, as shown. Angle the saw slightly and back-bevel the cut. This will ensure that the visible front edge of the coped baseboard will fit tightly against the adjoining baseboard.

Mitered lap joint

8d nails as needed

For the best looking corner on simple round-edged baseboard, a mitered lap joint is used. The board on the left is cut first, and nailed in place. The second board receives corresponding cuts.

Trimming Out the Main House

Building a Victorian headquarters for George Lucas

by Rick Wachs

In a rural valley just north of San Francisco, a working ranch is being developed. In this grassy, rolling countryside, dairies are common. But on this ranch, the old milking shed is where power tools and air compressors are repaired. Next to it, a new 15,000-sq. ft. barn houses a mill shop outfitted with the best woodworking tools that money can buy and staffed by some of the most talented woodworkers on the West Coast. Their job is to produce parts for the various buildings that will be a complete pre-production and post-production filmmaking complex for Lucasfilm. The ranch will eventually have facilities for film editing and sound recording, as well as offices for the people who will maintain the complex. The buildings are arranged in small clusters across 2,600 acres of land, and each one is done in a turn-of-the-century style, such as Shingle or Queen Anne Victorian.

The jewel of the ranch is the Main House (photo facing page, top), and it stands at the upper end of a small valley. The Main House is a grand Victorian flagship of a building with 6,000 sq. ft. of covered porches, a turret, a tower, dormers, gables with fretwork panels, and numerous chimneys. One wing has raised elliptical panels above banks of casement sash. The tower with arch-top sash is decorated with starburst panels, corbels, dentils and built-up cornice moldings. Almost 2,000 ft. of ornate handrail encloses the porches. The house has almost 500 windows, many of which are double-hung sash with radiused glass.

The Main House covers 45,000 sq. ft. on four levels. It includes executive offices, a library, a screening room, a commercial kitchen, breakfast, dining, conference and living rooms, ancillary offices and editing workrooms, not to mention all the bathrooms, coffee kitchens, mechanical and electrical service rooms, and an elevator. There are also 13 fireplaces and a 12,000-amp service panel.

The interior of the house overflows with exquisite woodwork. One cannot help but be awed by the luxuriant atmosphere. The construction took three years. Twenty-five finish carpenters, myself among them, worked on the house at the peak of the project.

Background—Because of the highly specialized functions of many rooms and because it was built to commercial code in earthquake country, the house embodies unusual structural, mechanical and electrical systems.

The framing is of 2x6s and 3x6s with tubular steel posts, K-braces, wide-flange beams and glue-laminated beams. Within the walls and floors are conduits carrying fiber optics, computer, telephone, security, stereo, heat and smoke-detection systems, fire sprinklers and HVAC ducting, in addition to plumbing and electrical systems. Because Lucasfilm uses the most advanced technologies, we even installed runs of empty conduit for linking electrical systems that haven't yet been invented. All walls have at least two layers of ⅝-in. gypsum board on each side, some as many as four.

Review and modification were common, as were frequent visits from the client. Collaboration was constant among the many designers, architects, millworkers and carpenters, all of whom were based on site. Given the client's substantial resources and his involvement in the visual arts, the design team was asked to do drawings on a scale reminiscent of those done in the days of the Ecole des Beaux-Arts. The massive roll of blueprints that stretched across my desk included 30-in. by 40-in. pages with drawings at ½ in. to a foot showing every piece of trim on a wall 25 ft. high. Drawings of this intricacy were done for all the details in the building, and set the tone for the millworkers and trim carpenters who had to turn them into reality. There are not many buildings going up these days with 7½-in. wide Victorian casing, radiused to boot. All the carpenters had to develop and perfect skills they had seldom used. Patience and a thoughtful approach were essential.

Redwood trim stock—Originally the house was to be trimmed entirely of oak, but plans changed when we found a local source of salvaged redwood timbers from first-growth trees. The beams, which were 6x16s, 20 ft. long, had been part of a bridge across the Newport Beach estuary in Southern California. The availability of the clear-heart redwood, which is a lot easier to work than oak, was the deciding factor.

For the Main House trim, we purchased 200,000 board feet of redwood in beams. The deep bluish-red color and tight vertical grain is unequalled in lumber currently available. Upon their arrival at the ranch, the beams were inspected and stacked. Those with vertical grain on the 16-in. face were marked as such, and set aside for door and wainscot panels. The re-

Crown moldings and balconies curve around the room as wainscot and stairway ascend toward the 20-ft. ceiling in the Main House lobby. These embellishments are in the formal style, one of three major trim styles used in the building. Every wooden element in the room, from the inlaid flooring to the compound curved handrails and wainscot panels, was made in the mill shop on site.

At the heart of Skywalker Ranch is the Main House. Since his early days as a film student at the University of Southern California, George Lucas has valued the steadying influence of working in a homelike atmosphere. He is now carrying out that principle on a grand scale—after three years of construction, this 45,000-sq. ft. Victorian Revival mansion is about to become headquarters for Lucas' pre-production and post-production filmmaking facilities.

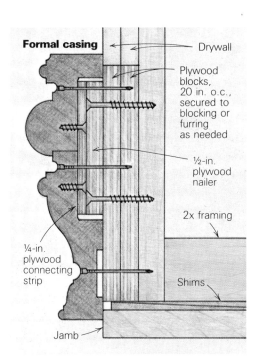

Formal casing — Drywall

Plywood blocks, 20 in. o.c., secured to blocking or furring as needed

½-in. plywood nailer

2x framing

¼-in. plywood connecting strip

Shims

Jamb

To anchor the formal casings, plywood blocks, seen on the right edge of the door in the photo below, were screwed to the stud wall and the drywall was notched around them. A ½-in. plywood nailer was then screwed to the blocks to provide solid backing for the 7½-in. wide door casings, as shown in the drawing at right.

Formal crown molding was designed to split into two separate crowns where it makes its turn into the lobby. At top, the radiused crown awaits a butt-joined section of straight crown to tie it into the turned drum. Note the blocks screwed to the wall to secure the crown. The finished installation (above) shows how the various trim elements come together, a major exercise in scribe-fitting where the crowns meet the drum.

Skywalker trim styles

Moldings and casings in the Main House range from the sumptuously ornate in the most public rooms to simple and serviceable in the more private workrooms deep within the building. The most elaborate trims are the formal and the Federal, followed by the Mission styles used in the library and screening room. A variation on the Mission style is Lucas trim, used in the filmmaker's writing room. Falling between the ornate and the strictly utilitarian are the semi-formal and informal offshoots of the formal style. These are used in public rooms, such as the breakfast room, enclosed porches and coffee kitchens. Below are drawings of the most frequently used trims in the house, showing sections of crowns, window stools and baseboards.

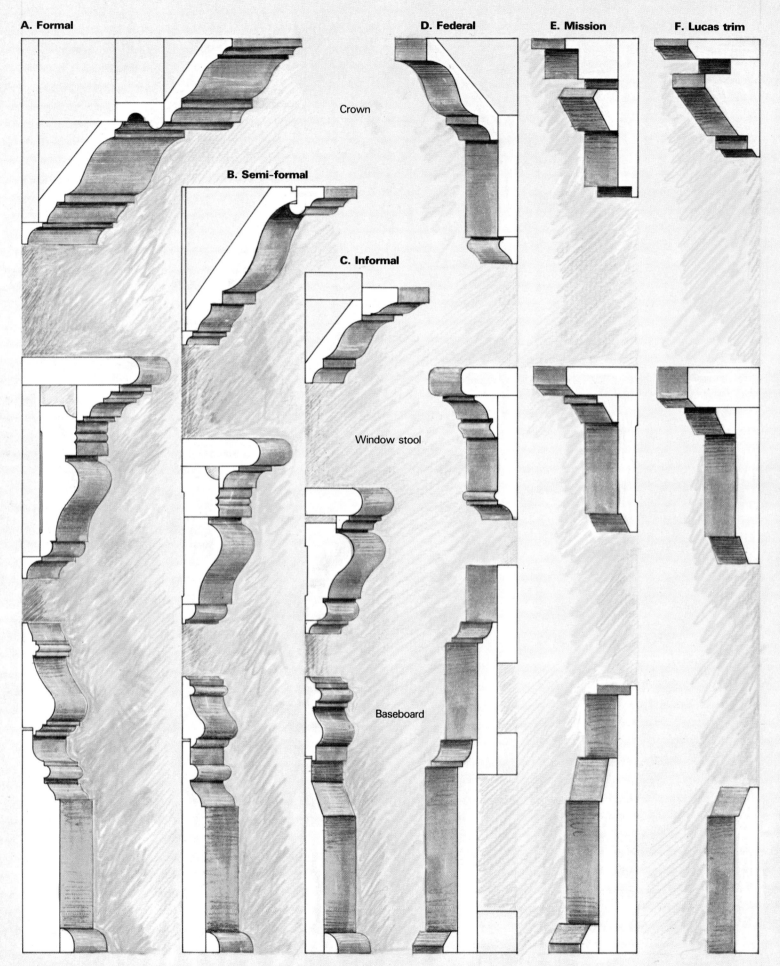

A. Formal

B. Semi-formal

C. Informal

D. Federal

E. Mission

F. Lucas trim

Crown

Window stool

Baseboard

Lucas' writing room is detailed in Lucas trim, a style reminiscent of Mission and Arts and Crafts motifs, with wide bevels and crisp chamfers. The gap in the crown molding is both a revealed shadow line and, in places, a duct for the cold-air return. The brass lamps were made by a Skywalker company set up solely to manufacture lighting fixtures for the Ranch.

The conference room of the Main House is done in Federal-style trim, which is characterized by stepped-out baseboards and fluted casings. The wainscot panels that surround the fireplace are redwood burl, framed by strips of ebony, Brazilian rosewood and brass. The hearth and fireplace surround are of green marble.

maining beams were resawn for door and window casings, crown and base, and a couple of hundred moldings with specialized profiles.

Trim styles—The principal trim styles in the house, which are shown in the drawing on the facing page, vary in degree of formality. The lobby (bottom right photo, p. 129), living room and music room are finished with the most ornate and massive trim in the house, which we called formal trim (A in the drawing). Throughout the house, transitions occur leading to semiformal (B), informal (C) and Mission-style (E) areas. In some rooms Federal trim (D and photo above right) creates the mood.

Many of the hybrid trim styles draw from traditional patterns and were modified according to the ideas and fancy of the designers involved in the project. The best example may be what we called Lucas trim (F). Used solely in his writing room (photo above left), it has a chamfered, rectilinear look that recalls the Mission and Arts and Crafts periods.

During the design and development of the interior trim styles, we made full-scale mockups of all the major wall treatments. These were 4-ft. wide by 9-ft. high wall sections, built and finished to the exact specifications of each trim area. On the formal-trim mockup, the wainscot panels could be slid out and replaced by alternate designs. Lucas made his final decisions about the trim styles after he reviewed the mockups, and we often referred to them during construction. It was quite a sight to see a half-

dozen sections of immaculately detailed walls leaning against the wall of the conference room during these reviews.

Formal casing. Formal casing is 7½ in. wide and 2¼ in. thick at its highest point. In order to rough out the stock from the 6x16s economically and to ensure a clean run on the molding machine, formal casing was made in two parts. Once the material had been milled, our millshop foreman decided to modify it further. He realized that securing the large bead on the outer portion of the casing could require a 20d nail to reach the framing through two layers of gypboard. The casing parts were plowed out to form a groove ¾ in. deep. This was to accommodate a ¼-in. thick plywood connecting strip and a ½-in. thick nailer (drawing, p. 129), which in turn could be secured to plywood blocks eliminating the need for such large nails in the casing itself.

A carpenter with the task of mitering the formal casing would first join the two-piece casing with a ¼-in. strip of plywood, glue and screws. Then he would make the miter cut on a 16-in. power miter box and test-fit it with its neighboring piece. No matter how accurate the corresponding profile, small variations in the casings made it necessary to fit each joint meticulously. Chisels, gouges, files, and sandpaper of various grits were used to make the infinitesimal adjustments at the miters to ensure smooth transitions from adjoining casings.

Once the corners fit properly, ½-in. plywood nailers were screwed to the framing, and the

casings were glued and nailed to the jamb and the nailers. Formal plinths were made out of casing offcuts. Their outer edges were ripped off and replaced by deeper and wider moldings, which create a ⅛-in. reveal on top of the plinth.

Formal crown. Formal crown (drawing A, facing page) is built up from four moldings that are nailed to backing blocks designed to align and secure the trim. The area trimmed in this style contains twelve radiused outside corners and two radiused inside corners. The mill shop made all the radiused parts from the beam stock. Because of the jungle of conduit and ducting inside the walls and ceiling, backing wasn't always easy to find. To secure blocks for the crown, we first put ½-in. plywood strips up with glue and screws. The blocks were then glued and screwed to the plywood strips on 16-in. centers.

The formal crown molding in the lobby is made up of two equal halves, which allows it to divide and become two separate crowns at the balcony. At this intersection, the lower portion of the crown wraps around the radiused wall and dies into a turned drum (photos previous page, middle and bottom left).

Screening room—The screening room (photo next page) is finished with redwood boards milled from 18-ft. long barrel staves salvaged from wine vats. Years of marination intensified the wood's naturally reddish hue, and the acid in the wine left dark streaks in places. Whenever we sawed into the wood, the odor of wine filled

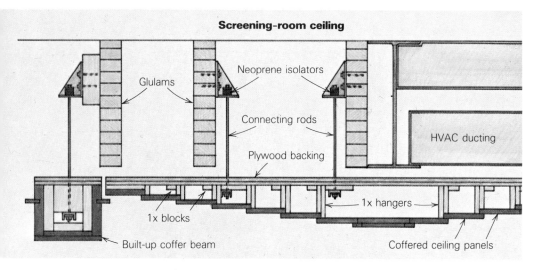

Screening-room ceiling

Glulams

Neoprene isolators

Connecting rods

HVAC ducting

Plywood backing

1x hangers

1x blocks

Built-up coffer beam

Coffered ceiling panels

the room. It was as if fine chefs instead of carpenters were at work.

The screening room is sound-isolated from the rest of the building. On the slab floor of the basement we poured a second slab. It is 4 in. thick and contains jacks made of threaded rods inserted into sleeves at 24 in. o. c. throughout. After the slab had sufficiently cured, a crew armed with wrenches turned the jacks in unison to the call of the lead man. The slab now rests on a series of points.

We built 2x6 walls upon the floating slab, and then suspended the ceiling from the glulams in the floor above the room. Neoprene isolators keep the ceiling and the walls from making any direct contact with the rest of the building. At the doors and projection ports, rubber gaskets separate the wood surfaces. In this vibration-free room there is an eerie silence. I was standing in the screening room during the Morgan Hill earthquake in 1983, and didn't feel the slightest jiggle, though it was felt just outside the room and sharply on the upper floors.

The trim material was sawn from the 3x6 staves in three layers. Boards from the first cut, closest to the inside of the barrel, were a deep red. The second and third cuts produced boards with less staining from the wine. In order to blend the three different colors in the room, the paint foreman used several red stains and Procion dye mixed with alcohol. With three coats of polyurethane finish, the woodwork glows.

The coffered ceiling in the screening room is attached to plywood hung from glulams. In each bay a series of 1x hangers on edge were glued and screwed to 1x blocks (drawing above left). Each of the wine-wood ceiling panels is finish-nailed to a pair of 1x hangers. The layout had to be precise because sprinkler heads are located at beam intersections and panel centers, lights are centered in smaller perimeter bays and the HVAC vents through slotted grilles around the center panel. Coved capitals join the pilasters to the coffer beams. They were made from band-sawn 1x sides with veneers forming the faces. Plywood was bandsawn to serve as backing for the 1x4 V-groove that fills out the cove between the capitals. The 1x4 paneling below the sill at the fabric panels is canted in at the base to limit sound reverberation across the room. Behind the fabric panels, perforated Masonite and insulation help absorb the sound.

Wainscot—The lower portions of the walls detailed in both Federal and formal styles are paneled with wainscoting. To accommodate the thickness of wainscoting and ensure straight walls, we furred the drywall above wainscot height with 1x4s and shims (drawing, facing page). This helped us to make up for the variations in thickness and inevitable warps in the framing lumber. Wainscot panels were delivered to the house in assembled sections. Measure-

Coffered ceilings and Mission-style trim work create a sumptuous atmosphere in the screening room, recreating the aura of Hollywood in the 1930s. When the projector rolls, there's room for 37 viewers in overstuffed chairs and couches covered in velvet.

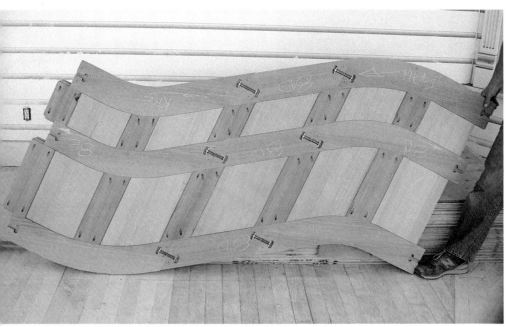

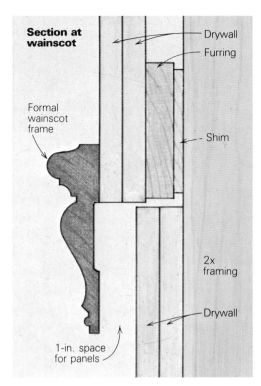

Section at wainscot

Drywall

Furring

Formal wainscot frame

Shim

2x framing

Drywall

1-in. space for panels

Wainscot frames were glued and screwed together from the back, then router-mortised for draw bolts. The individual panels were then glued together, with the draw bolts applying pressure as the glue set up. To make sure the wainscot casings would lie flat, the panel backings were shimmed into alignment (drawing, right). A finished panel installation can be seen in the photo on p. 128.

ments for the panel runs were taken in the field, and layouts were made in the shop. We used panels of various widths to fit each situation.

Panel units were put together with draw bolts and screws, as shown in the photo above. Adjoining sections were mortised for the draw bolts in the mill shop. In the field, we assembled all the parts of a panel run with glue and draw bolts and lifted the entire assembly into place. This way we got the best possible fits at the rail joints. Providing accurate backing was an extremely important part of the job. It had to be plumb, level and straight, and in alignment with the wall surface so the casings and wainscot cap would lie flat.

Coped joinery—There are thousands of coped joints in the Main House, from base shoe to crown and everywhere between. By far the most difficult cope cuts were those that fit to turnings, such as the one at the main stair (bottom left photo, p. 129). There was no way to precut these joints, and they had to be made by fitting and cutting. The material at the outermost part of the crown became very thin. With files and gouges, the backs were cut away until the profile of the crown fit properly to the turning.

Within the capacity of a 10-in. power miter box, moldings could be precut and then coped. On larger pieces such as semi-formal crown, the board was laid flat and compound-cut on the Rockwell (now Delta) Sawbuck. Depending on the shape of the back of the stock, it could be cut at about a 32° bevel with a 45° miter.

We set up one Sawbuck just to cut wide moldings. With a 40-tooth carbide-tipped blade, it made clean cuts, but we had constant trouble with the adjustable fence linkage. So we removed the fence and linkage, as well as the folding legs and wheels. We mounted what was left of the tool to a bench with extension tables, and then attached a rigid fence to the brackets on which the original fence had traveled.

Doors—A book could be written about the doors and hardware. All doors were made in our shop. There are fire doors, sound-gasketed doors, pocket doors up to 5 ft. by 9 ft., and many other kinds, too. We were licensed to manufacture twenty-minute-rated redwood panel doors. These have 2¼-in. thick rails and stiles, and ¾-in. panels glued back-to-back over ¼-in. Flametest, a fire-resistant material manufactured by the Masonite Co. (200 Mason Way, City of Industry, Calif. 91746). The panel, sticking and muntin designs match the different trim styles in the house. Doors that lead from one trim area to another have different patterns on each side. Some doors have four panels with informal trim profiles on one face, and two panels with formal profiles on the opposite side. Leading into Lucas' office, which is trimmed in oak, are two pairs of 3-ft. by 7½-ft. doors with redwood panel construction on one side and oak on the other. They came to be known as "redoak" doors. The doors from the conference room to the screened porch are redwood on one side and Douglas fir on the other.

We hung all the doors using a Stanley (now Bosch) router and template to mortise for hinges. All fire-rated doors have concealed closers mortised into the top rail. The mortises are lined with sheet metal so fire won't burn through the thin wood near the closers. For soundproofing, many doors have drop seals cut into the bottom rail. These spring-operated rubber gaskets press against the threshold when the door is closed.

Locksets are all box-mortise type. We used hinges with steeple pins, ball pins or flat pins, depending on the adjacent trim. One carpenter spent six months installing door hardware.

Library—The library was designed to look as if it had been built as an addition in the 1930s. The screening room is in a similar Mission style, as if it had been renovated at the same time. On the library wing, even the exterior siding is dif-

ferent to indicate its more recent place in the story. The library is bathed in the warm glow of amber light filtering through the 19-ft. dia. art-glass dome (top photo, next page) and through high windows around the mezzanine. All surfaces—bookcases, walls, coffered ceilings, moldings, headers and crowns—are redwood.

The plywood paneling for the walls and ceiling was made on site in our mill shop, where the 6-in. wide veneers from the beams were book-matched and laminated to plywood or particleboard cores. The shop people applied veneers to both sides to prevent warpage. In all, 200 sheets were made from 8 ft. to 10 ft. long by 40 in. wide. To square off the ceiling around the art-glass dome, four triangular panels with wedge-shaped veneers were scribe-fitted to match the radius of the ring beam that carries the dome. The beam is a tubular-steel compression ring faced with laminated curved boards on the sides and a bandsawn 2x on the bottom.

The library presented a tough layout job. Its design is based on a gridwork of radial lines, continuous from the ceiling to the windows to the pilasters on the mezzanine, through the railing to the ceiling under the mezzanine and to the battens and pilasters on the first floor.

A fireplace of dark red clinker brick with stepped corbeling is the focal point of the west wall. The brick was chosen because of its irregularities, which gave the hearth and facing a rustic texture. The mason carefully set the brick in uniform courses. He was particularly pleased with the way it looked. He cleaned up and went home. In the late afternoon the designer responsible for the library went out to inspect the fireplace. He picked up the mason's trowel and tapped a brick here, a brick there. To him it was too uniform. The following morning the mason returned. He was dumbfounded. Yet another example of collaboration.

Elegant in its singular deviation from the straight and angled is the spiral staircase. Built

entirely of wood, the staircase (photo bottom left) was milled and pre-assembled in the shop. The spiraling parts—stringer, rail and trim— were laminated on a 5-ft. dia. cylindrical form that looked like a great paddle wheel. With a 3-in. steel pipe as its axle, the form could be turned as boards were glued and clamped to it, and as parts were shaped. The stairway's center post was built out of staves glued around plywood circles with a 4x4 post extending through and projecting out the top to support the handrail. The underside of the stairway is paneled with 1x4s. The 1x4 compound-curve belly boards were laminated from ¼-in. thick strips bent over a form matching the structure of the stairs. The strips were glued and clamped in sets spanning the width of the stairs. Each piece was meticulously scraped and sanded to smooth the surface. The stair was finished in the mill shop, trucked to the Main House, and then twisted like a corkscrew into the library through a rough opening for one of the art-glass windows.

Library railing—To build the railing, we first framed a plywood curb between the rough 4x4 posts, which were bolted to the mezzanine rim joists. Meanwhile the shop crew made up the frame-and-panel post wraps. They are detailed with rabbet joints at the corners. In the top end, a 1x cap was inset for nailing into the post. Then we tacked shims to the bases of the 4xs to align all the post wraps and space them equally. When the wraps were slipped over the posts they were plumbed and the top 1x was nailed to its post.

Next we trimmed the curb between the posts. The top cap of the handrail is grooved for the bottom rail to nest in. All of the 2x2 baluster and rail material was milled, and the balusters were cut to length in the shop. On a router table, we dadoed the balusters to accept the rails. The length of each rail section varied slightly, so the rail pieces were different lengths in each bay. All of the 2x2 parts were put together on the bench except the mid-rails at the ends. These were left out to allow for screws into the posts. Once these sections were in place, the handrail was built up and set in place as a unit. Finally the post caps were assembled with mitered frames and nailed on.

Trimming out the Main House was a once-in-a-lifetime experience. That high standard of quality and access to the necessary expertise and tools made the job unique. The mill was right there to make all the parts, tailored to fit whatever circumstance required. The painters and finishers collaborated at each step to prepare, seal and finish the material. Their work ultimately gives the house its smooth, clean look. People from all departments were in constant communication. The designers and draftsmen, project managers and foremen, millworkers and carpenters all took part in seeing the job through. There was all the hustle and bustle of a large project, yet it came together with the kind of close teamwork that usually only happens on small crews. I'm still amazed by what we built. □

Rick Wachs was trim foreman on the Main House project. He is a member of Carpenter's Union Local #35, and lives in Mill Valley, Calif.

Many specialized studios were set up to supply the Main House with furnishings and hardware during its construction. Among them was the Skywalker Art Glass Studio, which crafted the lanterns and the hanging lamps, as well as the 19-ft. dia. leaded-glass dome that tops the Main House library (above). This view is from the mezzanine level.

The library mezzanine is reached by way of a spiral staircase (right), which was built around a hollow redwood column housing a 4x4 post that ties into the railing. The 1x4s sheathing the belly of the stair are made up of ¼-in. thick strips, laminated together on a form built to simulate the curvature of the stair. The treads were made of Honduras mahogany, which wears better than redwood.

Molding Character

Using a molder/planer on site to create a formal 18th-century interior

by Douglas Honychurch

As a Connecticut native and real-estate appraiser, I've gradually come to appreciate the examples of early Colonial architecture that remain in our area, and I particularly value the craftsmanship that made them worth preserving. My interest began with old houses that I noticed while driving through the New England countryside, and eventually led me to look for historic houses that were open to the public. I began buying books about Colonial houses and enjoyed studying their wonderful architectural details. Some of the books had cross-sectional drawings of walls, showing wainscoting and crown molds. These especially interested me.

The desire to do some of this work myself led me to purchase a Williams & Hussey molder/planer, and use it to run moldings for a remodeling project in my own house. But my real chance to design and mill Colonial trim came when my

brother asked me to help with a new house that he was building. I didn't need much coaxing. The house is a two-story Dutch Colonial with flared eaves, cedar clapboards and a corbeled end chimney (top right photo, next page). My brother gave me free rein in the design and construction of the architectural details.

Design—With a slate-faced fireplace on one wall and a large, multi-pane window on another, the 420-sq ft. living room was a designer's paradise. We decided to build twin bookcases on each side of the large window. A raised-panel wainscot would enclose the room, with a two-piece chair rail above and a two-piece baseboard below. The door and window casing would have a beaded edge toward the opening and a small band mold around the outside. On the doors, this casing would die into a con-

toured plinth block at the floor. The formal mantel surrounding the fireplace would be flanked by fluted pilasters (photo below). There would also be an elaborate crown mold all the way around the room. For all this work, we wanted to forego stock moldings and make our own trim wherever possible.

The designs came primarily from three books: *Southern Interiors of Charleston, South Carolina* by Samuel and Narcissa Chamberlain (Hastings House Publishers, 1956), *Architectural Treasures of Early America,* vols. 3 and 7 (reprinted in 1977 by Arno Press Inc.) and *Early Domestic Architecture of Connecticut* by J. Frederick Kelly (Dover Publications, 1963). The first two books are out of print (I found them in used-book stores), but the third is still available.

My brother and I wanted to create a room that felt like the late 18th century. Decisions

about what moldings to use where were based on my taste. I liked the wainscoting and fireplace designs in *Architectural Treasures of Early America.* But they were too elaborate. We simplified the moldings and eliminated the carvings. These decisions were also influenced by the limitations of my molder/planer, which can cut only ¾ in. deep. And of course, trial and error played a part in the design—tack on a molding, stand back and look—a process that led to many changes.

Construction—Unfortunately, the carpenter who had done such a fine job framing the house was behind schedule for his next job and couldn't be persuaded to do the complex trim work on the first floor. Our search for an equally skilled carpenter took some time, but on the recommendation of our painter, we met Ed Rockwell and looked at some of his recent work. We soon learned how fortunate we were to find him. His experience with this type of finish carpentry was extensive, since his family had been in the business for several generations. Unlike some tradesmen who are steeped in traditional lore, Rockwell was willing to try new designs and methods.

With our carpenter on the job, we were ready to make the moldings. The molder/planer we used was a Williams & Hussey model W7S (photo right), with power infeed and outfeed rollers (Williams & Hussey Machine Co., Elm St., Dept. 1361P, Milford, N. H. 03055). When I got the machine in 1984, I also bought a 2-hp motor to go with it. The whole outfit cost about $800.

We set up the molder/planer in the living room alongside Ed Rockwell's table saw, which was mounted on a rolling stand with a plywood extension added to the back of the table, both of which I found quite helpful. We used kiln-dried, select eastern white pine from New Hampshire. Ultimately we made seven different moldings (drawing, bottom right) and used three of them in the living room. The first step was to rip the boards to width on the table saw. We used a high-quality carbide-tipped blade, which made cleaning up the edges unnecessary.

Williams & Hussey offers a range of molding knives in standard patterns, and the company will also custom-grind special patterns. All you have to do is send them a drawing or a sample of the molding you want to cut. The cost of the knives is based on the overall width of the high-speed steel blanks (from 1 in. to 7 in.), not on the profile of the molding. I had custom sets of knives made, and they ranged in price from $72 to $120 per set.

To install these knives (two per set) on the molder/planer required only tightening some bolts. The alignment and registration of the knives were perfect without any adjustment. We used two plywood fences on the molder/planer, and secured them with small C-clamps that come with the machine. The knives had to be lowered into position and the fences adjusted and positioned so that the dimensioned boards would run through without any slop. We put the board between the two fences and lowered the knives to see where they would hit. If the alignment was right, we next adjusted the tension on the

Surrounded by fluted pilasters and a pair of raised panels, the fireplace mantel (photos above left and previous page) commands attention in a room full of architectural details. Both site-made and stock moldings were used for the trim work. To cut costs and avoid problems with wood movement due to moisture, the raised panels were made in one piece, using medium-density particleboard. The interior detailing, inspired by that of the 18th century, was a natural choice to complement the Dutch-Colonial styling of the house (above right).

Using a Williams & Hussey model W7S molder/planer on the site, Honychurch made many of the moldings used to trim the first-floor rooms. Molding profiles are shown in the drawing below.

Moldings made on site

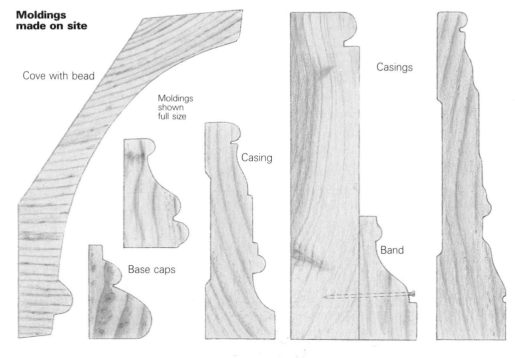

Cove with bead

Moldings shown full size

Casings

Casing

Base caps

Band

From *Fine Homebuilding* magazine (June 1987) 40:33-37

The mockup of the Greek-key fretwork on the mantel (above) was rejected because it would have been too complicated to make and apply. Saw kerfs were made in the back of the pine board to relieve cupping and were later covered by a thin band.

The crown mold in the living room (above) is just one 5-in. wide piece, though it looks built up from three pieces. It's a discontinued stock design called Curtis mold, and was ordered from a local mill. Around the fireplace (below), beaded cove mold was added under the crown.

feed rollers, which pressed the wood firmly onto the table while cutting, in addition to forcing it through the cutterhead. We learned how important feed-roller tension was when we broke a blade because the rollers weren't tensioned enough, and the stock began to chatter. Adjusting them was a real challenge—too little tension and the stock would flap violently and get chewed up; too much and it wouldn't feed.

Overall, the machine performed well for our needs and allowed us to make custom molding at less than custom prices. The blades stayed sharp; we used one set to make door and window casing for the entire house. These did need sharpening at one point, for which Williams & Hussey charged $15. To keep the knives sharp, it was important to prevent resin, pitch and sap from building up on them. We did this by carefully scraping them with an old chisel.

The fireplace—In the center of the 30-ft. long exterior wall is the fireplace. We wanted it to be the focal point of the living room. So one of our first decisions was to extend the 8-ft. section of wall that encompassed the fireplace into the room about 4 in., thereby breaking up the long expanse. This 8-ft. section is flanked by fluted pilasters, which we made on the job using a router and a fluting bit. The area between the pilasters is finished with an elaborate mantel and two large raised panels above it.

Our mantel design was suggested by one that we found in *Architectural Treasures of Early America*, vol. 7, and was a popular style in the eastern United States during the late 1700s. In deference to cost, we simplified the design by omitting the carved elements and using stock moldings to supplement the ones we made ourselves. We built the mantel directly onto the wall, furring out the main section with 2x4s nailed flat and covered with ¾-in. plywood to provide plenty of nailing surface.

At one point in the course of my trial-and-error design for the mantel, I considered using a Greek-key fretwork under the mantel shelf. I experimented on my table saw, cutting alternating grooves in a 1-in. piece of pine (photo top left), but realized that running it around the various offsets on the mantel would be tough. At this point, a trim carpenter who was working upstairs walked by and saw what I was trying to do. He didn't like the molding I'd made or how difficult it would be to apply, even though it wouldn't even be his job. He told me "You really know how to try a man's patience." He then built a jig for his router and made an alternative dentil molding that was good-looking and simpler to apply. It's the one we used.

Rockwell suggested that we embellish the crown mold in this 8-ft. section, and so we added a beaded cove at the bottom (photo bottom left). Also, at the extreme outside corners of this section, rather than a butt or mitered joint, we used a nearly circular beaded joint. Altogether, it took Rockwell and me a full week to complete the mantel.

Wainscoting—We decided to run a raised-panel wainscot around the entire living room, dining room and hall. The 34-in. chair-rail height

Photos at top of page: Douglas Honychurch

After some debate over whether the chair rail should protrude beyond the casing, the detail above was chosen as the appropriate way to end the molding at doors and windows.

was worked out to accommodate panels of pleasing proportion, separated by 3½-in. wide rails (horizontal border) and stiles (vertical border). In laying out each wall and determining panel sizes, we tried to keep the room as symmetrical as possible.

We chose not to use the traditional method of constructing raised panels—with stiles and rails mortised together, the raised panel let into a groove within them and the finished frames and panels applied to the wall. We wanted a method that would be less costly. Before sheetrocking, we had nailed plenty of blocking between all the 2x4 studs at the anticipated heights.

We started with the rails, nailing 1x6 pine around the room at chair-rail height. Then we ran the bottom rail so that once the 1x6 baseboard and base cap were installed, 3½ in. would be exposed above it. After the rails were up, we cut the stiles to fit between them. We used butt joints with a little glue and just nailed the pieces to the wall. No half-lapping was done. The pine wasn't a uniform thickness so we had to shim some of the joints to make them flush.

We then measured all the openings and calculated how much material we'd need for the raised panels. To save time and money, we decided to experiment with a medium-density particleboard called Medite (Medford Corp., Medite Division, P.O. Box 550, Medford, Ore. 97501) to make the raised panels. It is extremely dense and smooth and takes a beveled edge well, in addition to being very stable. Medite is available in 4x8 (actually 49-in. by 97-in.) and 5x8 (61-in. by 97-in.) sheets and a range of thicknesses from ¼ in to 1 in.

By using sheet goods for the panels, we were able to cut each one from a solid piece, as opposed to traditional panels, which are usually made from glued-up pine boards. Also, by using Medite, which is moisture resistant, we were able to reduce the chance that the panels would warp or check.

Using a table saw equipped with a carbide-tipped blade, we cut the Medite into panels to fit the 63 openings. We bought a carbide-tipped

Traditionally styled bookcases flank the large, multi-pane window. Before they were attached, all

raised-panel cutter, and arranged with a local woodworking shop to use their large shaper (1¼-in. spindle) to bevel the panels' edges.

We were extremely pleased with the way the raised panels turned out. We took the numbered panels back to the house and found that most of them fit without any need for adjustment because we had intentionally undersized them ¹⁄₁₆ in. all around. Then, using brads inserted with a brad driver, we ran ⅜-in. quarter-round around the openings to hold the panels in place and cover the gaps. We later discovered another advantage of our paneling method. When the electrician needed access at a particular spot behind

the paneling, we were able to remove only that one panel with ease.

The chair rail, instead of being flat on top, has a slight wave in it, which then steps down before protruding out in a rounded edge. Its downscaled profile resembles a traditional 18th-century chair rail. This was one of the few moldings we ordered from a local mill. We ripped the chair rail to a width of about 3 in., then scribed it to the wall with a pencil compass. We sat it directly on top of the 1x6 and nailed it into the studs. Underneath it, we ran the same band molding as on the door and window casings.

Rockwell and I disagreed about the treatment

the facings were beaded along their inside edges—a simple detail that adds depth and richness.

The same wainscoting design used in the living room also covers the walls of the foyer and runs up the stairs. But the crown molding in this area is a simple beaded design and was made on the site.

of the chair rail where it met the door and window casing. As I had designed it, the chair rail protruded about 1 in. beyond the casing. Rockwell strongly objected to this, believing that the profile of the chair rail should be shallow enough to die into the casing without sticking out. I insisted on my design and he relented, notching the chair rail so that it lapped onto the casing as much as it protruded beyond it. When he applied the band mold to the door casing, he decided to cope it around the chair rail. He did this 26 times throughout the house, and even he seemed pleased with the result (photo facing page, left).

Crown molding—The crown molding we used in the living room is actually one piece but looks like it was built up with at least three pieces (top right photo, p. 137). It is a discontinued stock design called Curtis molding and is made from a single piece of 5/4 by 5-in. pine. This molding has been made on the Williams & Hussey molder/planer, but we felt that it would be pushing the limits of the machine, and we therefore had it made for us at a mill. ·

For the crown molding in the center hall (photo above right) and dining room, we used a 4-in. beaded cove that I designed. This we did make on our machine. The profile required a

¾-in. deep cut, which the manufacturer says is the maximum. We made the molding in one pass, and it did test the machine's limits. We could hear the motor strain; in fact, the power demanded to cut this molding was so great that the lights dimmed when we ran it. A much simpler molding than the Curtis mold, it was easier to install, especially when coping the inside corners.

Bookcases—The bookshelves and cabinets that flank the multi-pane window (photo left) were designed by Rockwell, based on ones he'd done in the past. He built them in place, with 10-in. deep adjustable shelves above and 16-in. deep cabinets below. The upper portion of each bookcase has two bays made from 1x10 pine. Rockwell beaded the inside edge of the facings with his router. The exposed parts of the base were made with birch plywood, including the counter, which was then nosed with window stool ripped to 1¼ in. Under the nosing we ran the same band mold as below the chair rail. Rockwell made the raised-panel cabinet doors on site, using his router and table saw. The edges of the door openings are also beaded, which adds a satisfying depth to the design. □

Douglas Honychurch is a real-estate appraiser in Trumbull, Conn.

Raised-Panel Wainscot
Traditional results with table saw and router

by T. F. Smolen

Installing a traditional raised-panel wainscot is a good way to transform a nondescript room into a more formal space. It's also a handsome alternative to replastering old walls that have been damaged over the years by feet, furniture and children. A wood wainscot is more durable than plaster or gypboard, and it relieves the unbroken plane of the wall with the delicate array of shadow lines created by moldings and flat surfaces, as shown in the photo at the top of the facing page.

The term "wainscot" is often loosely applied to various paneling treatments that cover the lower part of a wall. The raised-panel wainscot shown here is a traditional style based on frame-and-panel construction. The frames consist of vertical members, called stiles, and horizontal members, called rails. They support panels whose beveled borders and raised fields give this particular style its name. Each panel rests in its frame with its grain running vertically. Panel width is limited by the width of your stock, unless you edge-join two or more boards together.

Beneath the bottom rail of the frame, a baseboard extends to the floor. At the top of the wainscot, a molding called a chair rail covers the joint between the top rail and the upper section of the wall.

The moldings that I used in making this wainscot can be bought at most lumberyards, and it's possible to make the raised panels and their frames with a table saw. I used a slot cutter and router table to groove the inner edges of my stiles and rails, but you could handle these as well on a table saw with a ¼-in. dado blade.

Panel design—The wainscot that I installed in the dining room of my late Victorian home is traditional in design. I wanted its top to be about 41 in. from the floor. This finished height would include an existing 7-in. baseboard that could be left in place, and a 4-in. wide chair-rail molding that would overlap the top rail of the raised-panel frame. With the stiles and rails 4 in. wide, the panels would show 22½ in. of height. Their actual height would be 23½ in., since ½ in. of the panel edges would be let into the grooved frame all around (drawing, above right).

The width of the raised panels was determined by the distances between corners and the door, window and cabinet frames in the room. In each wainscoted section, I wanted

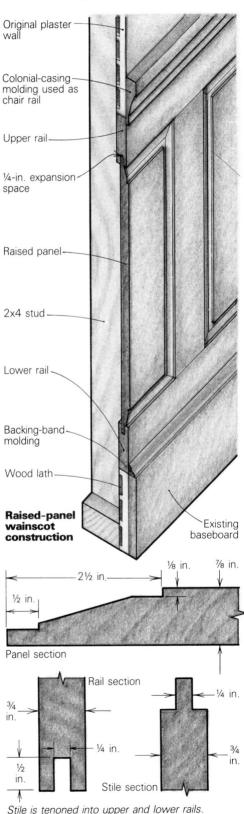

Original plaster wall

Colonial-casing molding used as chair rail

Upper rail

¼-in. expansion space

Raised panel

2x4 stud

Lower rail

Backing-band molding

Wood lath

Raised-panel wainscot construction

Existing baseboard

2½ in.

½ in.

⅛ in. ⅞ in.

Panel section

Rail section

¾ in.

¼ in.

¼ in.

½ in.

Stile section

¾ in.

Stile is tenoned into upper and lower rails.

the panel size to be uniform, so I divided each section to allow equal spacings between stiles. The largest section has four panels, each of which is 12¾ in. wide; the smallest section has a single 18-in. wide panel.

I milled the panel stock from 4/4 roughsawn pine boards about 14 in. wide. The wood, originally intended for flooring, was air dried. I had one side planed down to a thickness of ⅞ in., which allowed for a full ⅛ in. of relief on the raised panel and a sturdy ¼-in. thick tongue around the panel perimeter. Tight knots were acceptable, because I planned to paint the finished wainscot.

Raising panels—I set up a cutting schedule that included all panels for the wainscot, crosscutting the planks to 23½-in. lengths, then ripping them to finished width (the distance between inside edges of the stiles plus a ½-in. tongue allowance on each side). Then I made a template of the panel profile, which I set against the sawblade or dado head when setting up for a cut in order to produce the proper bevel and depth.

Making a raised panel with a profile like the one shown in the drawing below left requires three cuts on each side—one to form the bevel, one to form the tongue on the panel's edge and one to form the shoulder on the edge of the field.

To cut the bevel, I used a carbide-tipped combination blade because it produces a smooth surface that needs little sanding. To set up the saw for the bevel cut, I first set the arbor angle and fence distance to match the bevel on the template. Then I clamped a guide board to the tabletop, parallel to the fence and ⅞ in. away from it. This guide board aligned and steadied the on-edge workpiece as it was fed through the saw (photo facing page, center). Without it, you'd have a troublesome, hazardous time keeping the bevel straight and true. Even with the fence, the blade had to cut through just over 2½ in. of wood, so I held the stock securely and fed with slow, steady pressure. The next time I need a similar setup, I'll use a 4x4 as the auxiliary fence.

Because some boards cup slightly after they are surfaced, I found it best to make cross-grain bevel cuts soon after the boards had been cut to their finished sizes.

When the bevels had been cut on all four sides of the panel, I completed the panel pro-

From *Fine Homebuilding* magazine (February 1984) 19:50-53

file using 1-in. wide planer knives (flat across the top) mounted in a Sears molding head.

The second cut (photo bottom left) removes a triangular section of waste at the edge of the panel to create the ¼-in. thick tongue that fits into grooves in the frame. To allow for expansion and contraction of both frames and panels, I trimmed ⅛ in. from the top tongue of each panel and ³⁄₁₆ in. from each side tongue (wood expands more across the grain than along it). The inner edge of the frame would be grooved to a depth of ½ in. to provide expansion space at the top and sides of each panel. Allowing for play in the fit of each panel in its frame is necessary if the wainscot is to survive years of fluctuating humidity. Too tight a fit, and the panels are likely to check or bow out of their frames.

The third cut produces a ⅝-in. wide land at the juncture of bevel and field, and a ⅛-in. high shoulder where the field begins, 2½ in. from the panel edge. No auxiliary fence is required for this cut, but the main fence needs to be set up exactly right. Here again I use the template to get accurate settings for the fence and blade. The blade should just graze the wood surface.

As soon as the panels were cut, I prefin-

In a traditional raised-panel wainscot, a solid wood panel sits in a grooved frame. A chair-rail molding covers the joint between the top rail and the plaster. Overlapping the bottom rail, a baseboard extends to the floor, as shown in the drawing on the facing page.

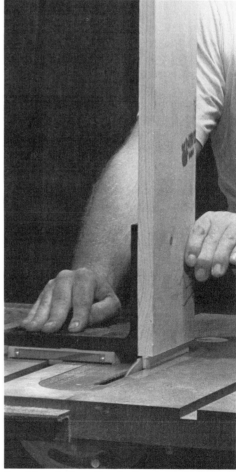

Table-saw setup. Top, a guide board, clamped parallel to the main fence, aligns the panel blank as the bevel is cut. Slow, firm feed and a carbide blade produce a smooth cut. For safety, let the blade stop spinning before the waste piece is removed. Above, a 1-in. wide planer blade cuts the tongue along the edge of the panel. At right, the author uses a tenon-cutting jig, attached to the saw's miter guide, to cut the tenons on a stile.

ished them to keep checking, cupping and wood movement to a minimum while I built the frames. I filled dings and small knots with plastic wood; then I sanded the exposed face with 80-grit paper and sealed knots with shellac to keep sap from bleeding through the finish coats of paint. Finally I gave each panel a coat of oil-base sealer compatible with the enamel finish I planned to use.

Stiles and rails—First I ripped 1x6 pine boards to 4-in. width. I cut all rails about 3 in. longer than their finished length so that after assembly I could scribe them for an exact fit to the walls on either side.

My design called for a ¼-in. wide by ½-in. deep mortise-and-tenon joint between stile and rail, so I grooved the inner edges of all the stiles and rails on my router table, using a ¼-in. slot cutter. Since the slot has to be cut down the exact center of the stock, I tested the setup on ¾-in. thick scrap before running stiles and rails through the machine.

To cut the tenons on the stiles, I used a thin-rim carbide blade on my table saw and a tenon-cutting jig on my miter gauge (photo previous page, bottom right). After a little touchup work with the chisel, I was ready to put frames and panels together.

Assembly and installation—I clamped the bottom rail of each panel section in the end vise of my workbench and fit each section together dry. Small pencil lines on panels and rails served as registration marks for centering each panel in its frame.

In assembling the wainscot, only the stile tenons get coated with glue. The panels are seated firmly in the frame, but not glued. This way, they can respond to changes in humidity and temperature without binding or bowing their frames. I used a small brush to spread glue on the tenons, and then assembled the frames and panels, snugging stile-to-rail joints together with pipe clamps until the glue set.

I wanted to nail the wainscot directly to the studs rather than installing it over the existing plaster, which was sound but presented quite an irregular surface. Leaving the original baseboard intact, I snapped a level line about ¼ in. above the installed level of the top rail and ripped out the plaster and lath. I scribed the plaster along this line with a utility knife, then ripped it off by hand. I cut the lath with a chisel and pried it off the studs with a hammer and small prybar.

Next, I nailed up the paneled sections, which I had purposely built slightly wider than the spaces they would occupy so they could be scribe-fitted to the walls. This left a slight gap between baseboard and bottom rail and between plaster and top rail. I used a 4-in. wide Colonial-casing molding at chair-rail height to cover the joint between the rough plaster and the top rail of the wainscot, and a 1⅝-in. wide backing-band molding at the baseboard and bottom-rail junction.   □

Ted Smolen practices law and does amateur woodworking in Danvers, Mass.

Modified wainscot: a raised panel with birch-veneer plywood and beveled molding

by Michael Volechenisky

A small ad in the local paper got me interested in building a raised-panel wainscot. It offered for sale "paneled wainscoting from a 150-year-old home" (wainscot is actually the correct term here). This would be just the thing for the dining room in my equally old house, which I was in the middle of redoing. But the age of the wainscot was unfortunately confirmed by its condition, and in any case there wasn't enough of it to go around my 15-ft. by 15-ft. dining room. So I decided to build my own from scratch, enlisting the advice of Luther Martin, a retired builder and woodworker who was sympathetic to the idea of recreating an old look with new materials.

Panel design—I planned to construct a frame-and-panel wainscot along traditional lines, bordering it with a baseboard along the floor line and a chair-rail molding along the top. But Martin didn't want to use solid wood panels because he'd seen long stretches of wood-paneled wainscot push walls out of plumb as a result of normal wood expansion. Cracking and cupping are other risks of solid wood panels. So I made my raised panels from ¾-in. thick lumber-core plywood with birch face veneers, and with slightly modified beveled molding. The molding, which is 2 in. wide, forms the beveled border of the panel, and the lumber-core plywood is the field. A tight tongue-and-groove glue joint between border and field and two coats of white enamel hide the fact that these panels weren't raised in the traditional fashion.

This alternative design has several advantages. First of all, the lumber-core plywood field is far more stable than its solid wood counterpart. Expansion and shrinkage are negligible, as are cracking and checking. And the birch face veneer is better than solid pine or fir if you're planning to paint—as I was—because it contains no knots or resin pockets, which could bleed through the finish.

Third, you can give the panel's bevel a fancier treatment than is possible with conventional techniques, since the border isn't an integral part of the field. The molding I used, for example, has a quirk bead at the inner edge of the bevel—an embellishment that suggests far more intricate work than was actually involved.

Michael Volechenisky lives in Sayre, Pa., and Pompano Beach, Fla. Photo by the author.

Making moldings—From top to bottom, my wainscot contains a chair-rail molding scribed to fit the plastered wall; two smaller moldings (a cove and a beaded stop) that fit over the frame and raised panels; and a baseboard scribe-fitted to the floor, with its top edge covered by a modified scotia molding (drawing, facing page). I made these moldings with my spindle shaper, but could have bought similar ones at the lumber store.

I milled the beveled molding for the panels that make up the wainscot on an old Hebert molder-planer. The Hebert, which is no longer made, is similar to the Williams & Hussey planer (Williams & Hussey Machine Co., Dept. 16, Milford, N. H. 03055), and both machines are shop-size versions of the larger, more powerful planer-molder machines used by lumber mills.

My machine has only one cutterhead, which is mounted horizontally above the table and holds a pair of knives. It is driven by a 1-hp motor, and is powerful enough to complete a molding in a single pass, providing you use knot-free wood that's not too dense. But to be on the safe side, I usually make my first cut to within ¹⁄₁₆ in. of the finished dimension and then run the stock through a second time to get a smooth surface that requires very little sanding.

Molder-planer manufacturers sell a variety of molding cutters to fit their machines, but I've often made my own from precision-ground tool steel (it comes in many sizes and thicknesses, and can be bought wherever metalworking tools are sold). After drilling two holes in each tool-steel blank so they can be bolted to the cutterhead, I transfer my molding outline to the blank and start removing metal. I hacksaw as much as I can, grind the shape to a 30° bevel and hand-file corners and coves where my bench grinder can't reach. After getting both cutters as nearly identical as I can, I mount them in the Hebert and run a trial piece of wood through. This tells me which cutter is doing most of the work by the flecks of wood that adhere to its cutting edges. More filing follows; then I hone the blades and install them. Since most of my molding runs are for 500 ft. or less, it doesn't seem necessary to harden my cutters.

All the moldings, stiles and rails for my wainscot were cut from basswood stock I'd been saving. Basswood works easily, and I've found that you can usually smooth it by hand with a cabinet scraper, with little or no sanding. It also takes a fine coat of paint, because it's knot-free and resin-free.

No matter what type of wood you use on a molder, you'll get better results if you make sure that each board you run through the machine has its grain oriented correctly relative to the cutterhead. To prevent small chips and tears in the molding, feed your boards so that their grain slants down toward the exit side of the machine.

I constructed the panels first. Each completed panel actually consists of five pieces—the birch-veneered lumber-core field and four bordering molding sections. As shown in the drawing, the tongue along the inner edge of the molding is designed to fit in a ¼-in. wide groove cut in the edges of each panel. I grooved the panel edges on my table saw, using dado blades. At the corners of the panel, adjacent molding sections are mitered. Once all the parts for a panel were cut and test-fitted, I glued them up.

After the panels were finished, I built their frames. Top and bottom rails for each 15-ft. side of the room were cut from 16-ft. long basswood boards. Using a hollow-chisel mortiser chucked in my drill press, I cut mortises in the rails to receive stile tenons. Then I grooved the inner edges of the frame on the table saw to receive the ¼-in. by ¼-in. tongue around each panel.

I test-fit the frame-and-panel sections for each side of the room, then glued and clamped them together. Before installing each section, I gave the back of the frames and panels two finish coats—the same number that the front of the wainscot would receive. Thus both sides of the wood can respond equally to temperature and humidity. Perhaps this wasn't necessary, but there hasn't been a single paint crack in the wainscot after eight years on the wall.

Though I could have installed the wainscot directly over the dining room's old plaster walls, they were in such bad shape that I stripped the room down to its studs and nailed up rock lath. Then new plaster was applied down to a temporary ground I nailed just below the height of the chair-rail molding that would top off the wainscot.

The frame-and-panel assembly was the first part of the wainscot to get nailed up. I made sure each wall section was level and used 8d finishing nails, positioning them close to the rail edges so they would be hidden by the covering layer of molding.

Once the four frame-and-panel sections were up, I added moldings to the rails. The baseboard and the scotia-style molding covering its top edge were mitered at the corners, as were the chair rail and its two adjacent moldings. A light sanding, followed by a primer coat and two coats of semi-gloss Kemglow enamel, finished the job. □

The gluing setup used to construct the panel consists of four bar clamps and a Formica-faced base slightly shorter and wider than the panel. At panel corners, the molding joint is a glued miter.

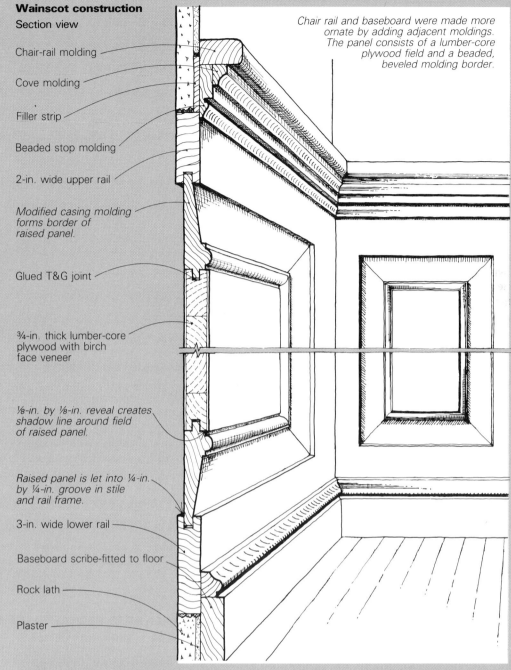

Wainscot construction
Section view

Chair-rail molding

Cove molding

Filler strip

Beaded stop molding

2-in. wide upper rail

Modified casing molding forms border of raised panel.

Glued T&G joint

¾-in. thick lumber-core plywood with birch face veneer

⅛-in. by ⅛-in. reveal creates shadow line around field of raised panel.

Raised panel is let into ¼-in. by ¼-in. groove in stile and rail frame.

3-in. wide lower rail

Baseboard scribe-fitted to floor

Rock lath

Plaster

Chair rail and baseboard were made more ornate by adding adjacent moldings. The panel consists of a lumber-core plywood field and a beaded, beveled molding border.

Gentle Stripping

The best advice is often "don't," but if you must remove paint, here's how

by Lowell Northrop

The first step in the process of removing paint from an old building is the same as the first step in any process of historical preservation: Hesitate. More often than not this boring and tedious job should be avoided. Analyze the existing layers of pigment (and document them, if the house is of historical significance); talk to experts and read as much as you can about the subject. Only proceed if you're convinced that proper preservation of the structure depends upon stripping its layers of paint. If, for example, water-based paints have been interlayered with oils, or films of dirt lurk between coats, or thick globs of pigment obliterate delicate detail, then you might be justified in going ahead. The best paint-stripping process is the one that is gentlest to the surface being painted. The worst method for wood is sandblasting—it tears up the wood and completely changes its surface character. Power sanding (and also power brushing) inevitably leads to scarred wood and can do terrible things to Victorian moldings. A good operator can do some tricky stuff with a power sander for about an hour a day, but he'll be tired for the other seven, when mistakes and ensuing damage are likely to occur. A problem with heat guns and torches is that they operate at temperatures above the kindling point of wood, about 500°F to 600°F, and can cause fires. When used on lead-based paints found on many older structures, they also create lethal fumes. Dry-scraping loose paint off a structure is only as good as the person doing the work. Conventional paint scrapers inevitably carve the wood more easily than they remove the paint.

I became aware of a fast, effective, gentle process for removing four or five layers of paint at a time from flat surfaces at a rate of about a square foot per minute through "Improved Process for Stripping Paint" (*The Old-House Journal*, Vol. 11, No. 8). In response to ordinances in some cities mandating that paint containing more than 1% lead be removed from all structures, the National Paint, Varnish and Lacquer Association, Washington, D.C. (now the National Paint and Coating Association) conducted research on methods of removing paint. They proved that using water-soluble paint strippers containing methylene chloride with the steam of a wallpaper steamer is a most effective method.

Methylene chloride dissolves the paint, and its solubility is increased by the heat of the steam. Also, the methylene chloride boils at about 104°F, causing the paint film to foam and bubble, which loosens it from the surface. Jasco Speedmatic, available from most hardware stores, is one good stripper; but make sure that the one you buy contains no benzene or carbon tetrachloride, both highly toxic.

Preliminaries—Choose the right time of year to do the work. The weather should be warm and dry, but not too hot. Very hot, dry weather and direct sunlight will harm the exposed wood, as will rain. Because steaming often increases the moisture content of the wood, you will need about a week of good weather for the wood to reach equilibrium with the atmosphere before repainting. This is a good time to tend to repairs.

Before beginning to strip your structure in earnest, experiment with the method on an inconspicuous area. This will verify that it is appropriate for your particular job. Methylene chloride can be absorbed through the skin, so make sure you wear protective eye and skin equipment, and know remedies for exposure, especially to eyes.

You'll need a map of action before you start so you won't miss areas, especially the inconspicuous ones. Measured drawings of elevations of the structure are useful for keeping track of who

Lowell Northrop, of Pacific Grove, Calif., is a preservation technologist.

+---+
| **Helpful Publications** |
| "The 8 Most Common Mistakes in Restoring |
| Historic Houses (and how to avoid them)," by |
| Morgan W. Phillips, *Yankee* magazine, Decem- |
| ber 1975. |
| *The Secretary of the Interior's Standards for* |
| *Historic Preservation Projects*, stock no. |
| 024-016-00105-2, $2.30. Available from the |
| Superintendent of Documents, U.S. Govern- |
| ment Printing Office, Washington, D.C. 20402. |
| (Enclose payment with all orders.) |
| *NIOSH Recommended Standards for Occupa-* |
| *tional Exposure to Methylene Chloride*, stock no. |
| 017-033-00194-4, $2.45, also from the Super- |
| intendent of Documents. |
+---+

is accountable for designated areas, if you're having others do the work for you. Proceed from the top of the structure down and finish each area before going on to the next. It's a good idea to establish a standard of workmanship (no missed spots, no carved-up wood, no lawn covered with scrapings) during the test-strip.

Place a protective covering over the ground, plants, electric and gas meters and phone connections. The substance removed is sticky, and unless you want to strip it again from something else, it should go onto a drop cloth or into a container. Plan for the removal of scrapings from the site—you will have an abundance of them.

Stripping the paint—To begin, apply a coating of stripper with a natural-bristle brush. Work with areas about 4 ft. square. If the stripper does not remain on the surface as a jelly for at least 60 seconds, recoat until it does. Wait 10 to 30 minutes, depending on the number of paint layers to be removed, and then apply steam from a wallpaper steamer for 10 seconds. Immediately remove all the leathery substance created. When using straight-edged scrapers, round the corners first to prevent cutting the wood. Be careful that no spots of paint remain after scraping.

For detail work where using the broad steamer pan is impractical, remove the pan and attach a compressed-air nozzle (available in most hardware stores) to the steamer hose. If the wood is sound and you follow the procedure, the remaining surface, with perhaps only some light sanding, should be paintable. If the whole structure has a residue, wash it with mild detergent and rinse it with water at low pressure. Although the water-soluble stripper-and-wallpaper steamer method is far superior to any other I've tried, stripping paint remains a tedious and risky job. I am researching other methods for paint removal, such as low-pressure air-blasting using flour as an abrasive, moisture migration (a radical procedure whereby a structure is encased in a plastic bag and exposed to direct sunlight while moisture is generated within), and an interesting improvement to the method discussed here using a nozzle I designed and high-pressure (500 psi) steam to remove paint softened by methylene chloride. I am in the process of patenting this nozzle. □

Period Moldings

A primer on these touchstones of Neo-Classical architecture

by Norman L. Vandal

Moldings are structurally non-essential building elements that help ease the transitions between large, primary structural elements. In Classical Greece and Rome, these primary elements were the plinth, the column, the capital, the entablature and the pediment (these and other architectural elements are explained in the Glossary on the next page). Over the years, Classical orders—the interrelationship of the dimension, proportion and location of these elements—were established. Composed of both structural and non-structural elements, they became accepted as proportionately correct and aesthetically pleasing. These strict proportions were adapted much later, when a maturing and increasingly humanistic Europe turned to the Classical past for architectural inspiration.

The Neo-Classical period lasted 150 years or so, and passed through several phases, known in the United States as Georgian (or Colonial), Federal and Greek Revival. There was no abrupt chronological dividing line between these styles, and in some cases overlap of styles is subtly or very apparent. The Classical forms were subject to various vernacular interpretations by country builders, many of whom were quick to improvise. A craftsman who owned planes for making Federal moldings wouldn't have been likely to discard these tools and get the new ones just because the Greek Revival style happened to be in fashion.

Nonetheless, each of the periods is characterized by the use of particular moldings to embellish essential architectural components. These moldings are distinctly different in each period. On pp. 147-149, the profiles of some of the moldings most characteristic of the different periods are drawn to scale (a profile is the combination of curved and straight parts that form a well-proportioned, graceful whole). They can help you date or restore period structures.

The Greeks and Romans carved their moldings in marble or stone, or cast them in aggregate. The inherent weaknesses in the stone were design determinants, and thus thin edges and steep projections were avoided. As a result, their moldings were often bold and bulky in section.

When Neo-Classical architecture began to catch on in late 17th-century England, however, wood was the most common material for residential building. All of these moldings were cut with wooden planes that were designed for specific profiles. Some of the simpler configurations were produced on the building site, but the larger, more elaborate ones (bed and cornice moldings and bolections, for example) required specialized planes and the expertise of the shop joiner to make them correct and consistent. Moldings were cut by hand this way until the middle of the 19th century. I do a great deal of restoration and reproduction work, and I still make and use such planes.

Here's a short primer on the characteristics of the three periods.

Georgian period (c. 1720 to 1790)—At this time, designers and builders in England were abandoning the motifs of the Jacobean period,

Georgian entrances were often elaborate, formal and robust. Builders imitated Roman moldings, and based their details on segments of the circle.

Federal details were still based on sections of the circle, but they were lighter and more delicate. Windows had thinner mullions and, often, semicircular tops.

Greek Revival detailing was based on the ellipse. Architects and builders consciously turned to the cradle of democracy as an appropriate model for American architecture. Columns, pilasters and moldings were larger, but simpler. Facades became grand, often harking back to the Parthenon and other Greek temples.

Glossary

Architectural elements

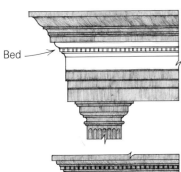

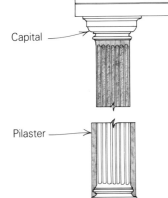

Pediment: The triangular space that forms the gable end of a peaked roof.

Entablature: The horizontal portion of a structure, which is supported by the columns. The entablature, from bottom to top, is composed of the architrave, the frieze, the cornice, and, in some interpretations, the pediment.

Cornice: Outside, the uppermost decoration on a structure, found either at the top edge of the pediment or at the top edge of the entablature where a pediment is not present. Inside, the molding at the intersection of wall and ceiling.

Fascia: The vertical face of the projecting cornice. The cornice molding is applied to the fascia.

Soffit: The horizontal underside formed by the projecting cornice as it overhangs the frieze.

Bed: A molded decoration at the intersection of the vertical frieze and the horizontal soffit. In profile a bed molding is similar to or the same as the capital.

Frieze: The portion of the entablature directly below the soffit. At the top edge of the frieze, below the soffit, is the *bed molding.*

Architrave: Outside, the lowest portion of the entablature, directly above the capital or the top of the columns. The moldings that decorate the architrave are often repeated on interior and exterior window and door casings, and these are also called architraves.

Capital: The molded decoration found at the top of a column or pilaster. It softens the transition between the vertical column and the horizontal entablature.

Pilaster: A vertical element made to resemble a column partly set into the wall.

Plinth: The block that the architrave or column sits on.

Chair rail: A molding running around a room at the height of the back chair posts, probably introduced to protect wall surfaces from being marred by furniture, but clearly accepted as a decorative element.

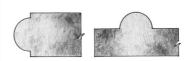

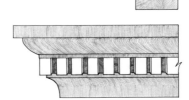

Wainscot: An interior wall treatment using boards or panels to cover the wall from floor to about window-sill height. Wainscot can also be a much broader term used to describe a manner in which boarding is used in various applications, including the construction of a particular form of furniture.

Moldings

Astragal: A convex, semicircular molding—usually applied—which projects above the surface of a flat plane.

Bead: A small, rounded molding usually found at the edge of a board. It is usually planed or carved, not applied. The most common architectural molding.

Quirk bead: A bead that has a narrow groove along one edge, and so appears to be separate from the surface upon which it is planed. Other moldings, such as ogees, can also be quirked.

Thumbnail bead: A molding in the form of a quarter-round, planed at the edge of a board with a slight step down from the surface upon which it is cut. Usually found on the rails and stiles of Georgian doors and fielded panel walls.

Bolection: A profile or group of moldings that separates two planes and projects from the surface of both. Usually found surrounding Georgian fireplaces.

Cove: A rounded, concave molding, sometimes called a *scotia.*

Dentil: A small, rectangular block in a series that project like teeth. Dentils are usually found as elements in a cornice, and are thought to represent purlins projecting beyond rafters.

Ogee: A molding that is formed by a continuous double curve, concave below, convex above. Sometimes called *cyma reversa.*

Reverse ogee: Also called *cyma recta.* An S-shaped molding convex below, concave above.

Ovolo: A convex molding—a quarter circle in Roman architecture, but a more elliptical curve in the Greek—which steps down from the surface on which it's planed and has a step at the bottom end of the curve.

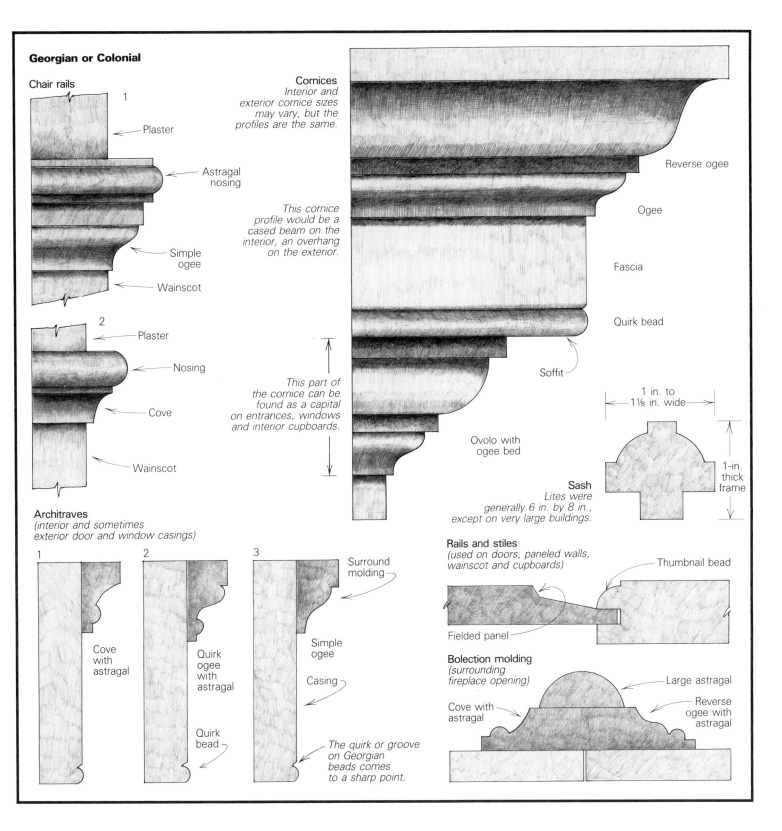

Georgian or Colonial

Chair rails

1
- Plaster
- Astragal nosing
- Simple ogee
- Wainscot

2
- Plaster
- Nosing
- Cove
- Wainscot

Cornices
Interior and exterior cornice sizes may vary, but the profiles are the same.

This cornice profile would be a cased beam on the interior, an overhang on the exterior.

This part of the cornice can be found as a capital on entrances, windows and interior cupboards.

- Reverse ogee
- Ogee
- Fascia
- Quirk bead
- Soffit
- Ovolo with ogee bed

Sash
Lites were generally 6 in. by 8 in., except on very large buildings.

1 in. to 1⅛ in. wide

1-in. thick frame

Architraves
(interior and sometimes exterior door and window casings)

1
- Cove with astragal

2
- Quirk ogee with astragal
- Quirk bead

3
- Simple ogee
- Casing

- Surround molding

The quirk or groove on Georgian beads comes to a sharp point.

Rails and stiles
(used on doors, paneled walls, wainscot and cupboards)

- Thumbnail bead
- Fielded panel

Bolection molding
(surrounding fireplace opening)

- Cove with astragal
- Large astragal
- Reverse ogee with astragal

which was characterized by the use of stone and masonry in early attempts to imitate the Classical forms. Wooden houses began to replace stone, and this led to more refined Classical lines. Guidebooks were published in England which heralded the new style, called Georgian after the four Hanoverian King Georges whose reigns began in in 1714. The trend crossed the Atlantic and took hold in the increasingly prosperous Colonies, where builders were quick to abandon the older, almost medieval styles.

Georgian buildings were larger and more symmetrical than their predecessors. Elaborate entrances that resembled scaled-down Classical temples were composed of pilasters, entablatures and ornate pediments. The larger windows were treated with capitals or cornices. Bed and cornice moldings were applied to soffits and fascias. The overall impression was massive, formal and ornate.

Inside, the austerity of the Pilgrim-century house gave way to rich ornamentation, and moldings became an important design element. Posts and girts, formerly left exposed, were cased with pine. Ceilings were plastered. Paneled walls and wainscot came into vogue, along with appropriate Classical moldings. The fireplace wall became a focal point, with the opening surrounded by a large molding called a bolection. Cornices at the intersection of wall and ceiling were the crowning touch.

The moldings of this period were bold and heavy. Their curves, like those of the Roman moldings they imitated, were based on segments of the circle. American builders interpreted the Classical style literally, and the molding profiles were not really elegant or refined. But this period did signal the acceptance of moldings as necessary elements in architectural ornamentation.

Federal period (c. 1790 to 1825)—This post-Revolution style was also spawned in England, where it is called Adamesque, after

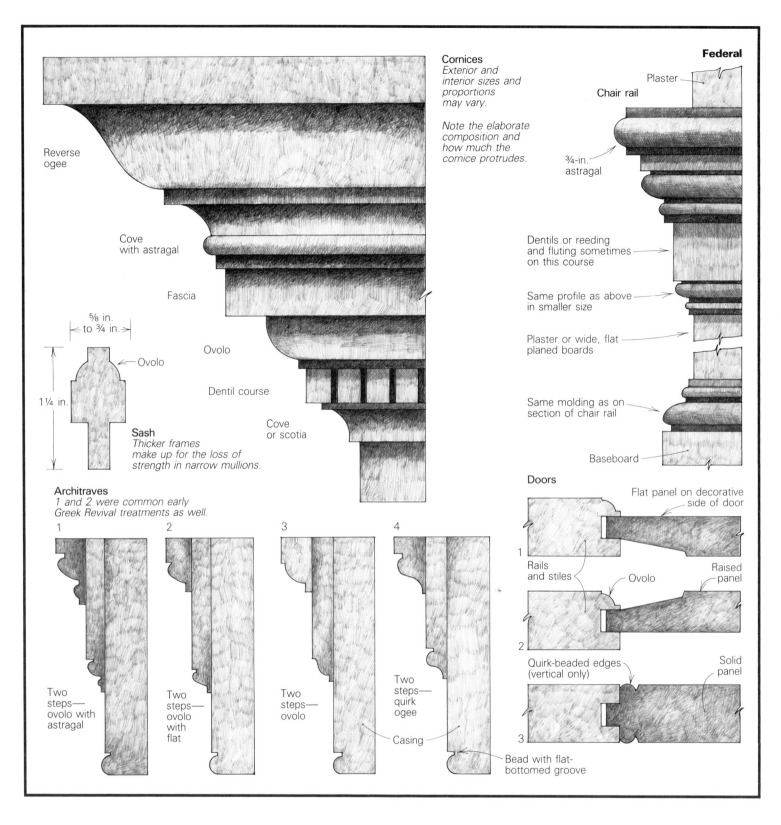

Cornices
Exterior and interior sizes and proportions may vary.

Note the elaborate composition and how much the cornice protrudes.

Reverse ogee

Cove with astragal

Fascia

⅝ in. to ¾ in.

Ovolo

Ovolo

1¼ in.

Dentil course

Sash
Thicker frames make up for the loss of strength in narrow mullions.

Cove or scotia

Architraves
1 and 2 were common early Greek Revival treatments as well.

1

2

3

4

Two steps— ovolo with astragal

Two steps— ovolo with flat

Two steps— ovolo

Two steps— quirk ogee

Casing

Bead with flat-bottomed groove

Federal

Plaster

Chair rail

¾-in. astragal

Dentils or reeding and fluting sometimes on this course

Same profile as above in smaller size

Plaster or wide, flat planed boards

Same molding as on section of chair rail

Baseboard

Doors

Flat panel on decorative side of door

1

Rails and stiles

Ovolo

Raised panel

2

Quirk-beaded edges (vertical only)

Solid panel

3

---

the brothers, Robert and James Adam. Boston architect Charles Bulfinch brought the new forms back from England and used them in several noted buildings, among them the Massachusetts Capital. Asher Benjamin, a student of Bulfinch's, heralded the new Federal style when he published his builder's guide, *The Country Builder's Assistant,* in 1797.

During this period, the Classical models in molding ornamentation were refined. Joiners took advantage of the fact that wood could be worked to yield thinner edges and flatter projections. Lightness and delicacy became the new guidelines of design.

Buildings were given a lighter and airier

feeling. Windows got bigger again. The low, squat appearance of Georgian structures was replaced by a sense of verticality.

The larger window panes had thinner mullion profiles. Federal entrances were reduced in scale, and semicircular gable-end windows became popular.

Inside Federal-style houses, mantelpieces, often with pilasters and carved friezes, became focal points in formal rooms. Plastered surfaces replaced paneling in many parts of the house. Wainscot gradually disappeared, leaving only the molded chair rail and the baseboard. The interior cornice was decorated but lightened. The large expanses of plas-

ter served to set off the lighter and more delicate moldings, and expressed their new importance. Moldings were meant to be noticed and appreciated.

Greek Revival period (c. 1820 to 1840)— This was a time of conscious return to Greek forms, which were considered to be purer than the Roman forms used in earlier periods, and more suitable for the architecture of a young republic. The Greek differs from the Roman in that all parts in the order are larger, and convey a sense of solidity and simplicity. There are fewer ornamental members than in the Roman, which on large structures can be

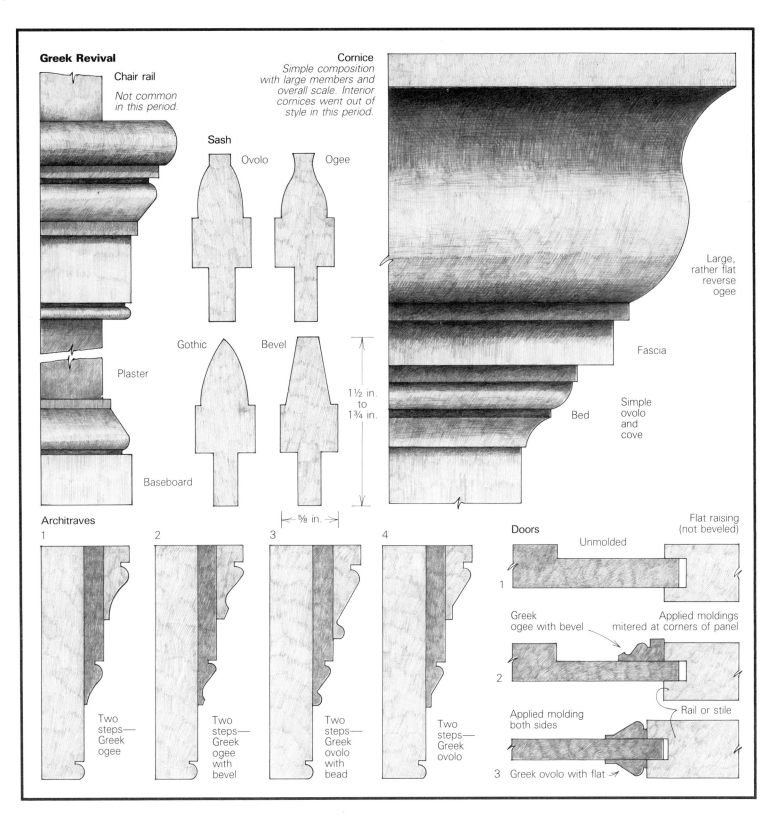

Greek Revival

Chair rail

Not common in this period.

Cornice

Simple composition with large members and overall scale. Interior cornices went out of style in this period.

Sash

Ovolo

Ogee

Gothic

Bevel

Plaster

1½ in. to 1¾ in.

⅝ in.

Baseboard

Large, rather flat reverse ogee

Fascia

Simple ovolo and cove

Bed

Architraves

1

2

3

4

Two steps—Greek ogee

Two steps—Greek ogee with bevel

Two steps—Greek ovolo with bead

Two steps—Greek ovolo

Doors

Flat raising (not beveled)

Unmolded

1

Greek ogee with bevel

Applied moldings mitered at corners of panel

2

Applied molding both sides

Rail or stile

3 Greek ovolo with flat

less confusing. The entablature is larger, with more room for ornamentation.

Roman molding profiles are composed from segments of a circle; Greek moldings from segments of an ellipse. During the Greek Revival period, it was believed that the flatter, elliptically shaped moldings offered a more pleasing reflection of light from their surfaces. The rounder Federal moldings began to fall into disuse. In some rural interpretations of the Greek Revival style, flat, unmolded stock was substituted for moldings, and the effect was quite pleasing.

The new Greek Revival style was a marked departure from the Federal period, and Asher Benjamin kept pace with the times. The sixth edition of his new guide, *The American Builder's Companion* (1827), presented drawings of the Greek orders for the first time, and the impact was tremendous.

On a Greek Revival exterior, the larger proportions of columns and pilasters, the wider entablatures and the larger yet simpler pediments and cornices give Greek Revival structures a solid appearance reminiscent of ancient Greek temples like the Parthenon. The gable end, turned to face the street, became the most important facade. Elliptical sash supplanted the Federal semicircle over entrances and in gable ends. Pedimented entrances lost

popularity, and sidelites were used instead of a transom above the door.

The biggest change inside was that the fireplace was replaced by the more efficient woodstove. As a result, the mantel nearly disappeared. Interior cornices were deleted, as were chair rails. A movement to elevate the staircase as the focal point, which had its roots in the Federal period, culminated in the Greek Revival period with the design and execution of the free-standing elliptical staircase, a marvel of Neo-Classical architecture. □

Norman Vandal makes period architectural components and period furniture in Roxbury, Vt.

Table-Saw Molding

The secret is in the order of cuts

by Bruce Andrews

When the landmark Winooski Block was finished in 1862, the builders festooned it with all manner of ornamental moldings and wooden filigree. But by the time we (Moose Creek Restorations Ltd.) got the repair contract, 117 Vermont winters had weathered, cracked and split all of its remaining woodwork. Three-fourths of the building's cornice moldings were either rotten or missing. We were to replace 10,000 linear feet of various moldings, not one of them a type manufactured today, and we didn't even own the usual tool for milling moldings, the spindle shaper. We still were able to complete the job, relying on our table saw and a lot of careful planning. We found that the table saw could handle most any profile—it could even scoop out concave curves—but we also learned that every profile required its own sequence of cuts. Figuring out that sequence is the heart of our method.

The first thing we worried about was getting enough good stock. Molding stock must be the highest quality, close grained and knot free. We were still short of stock after several deals to obtain a couple thousand board feet of Vermont pine in varying widths, thicknesses and lengths— all rough cut and in need of finish planing, dimensioning, and in some cases, drying. We were bemoaning our plight when two young entrepreneurs wandered into our office. They asked if we knew anyone who could use several thousand board feet of redwood and cypress beer-vat staves from the old Rheingold Beer brewery that was being dismantled in Brooklyn, N.Y. Well, yes, we probably knew someone. The wood reeked of stale beer, but it was superb for our purposes. It was straight, close grained and of course, well seasoned.

Before any shaping could be done, we had to prepare our stock. We thought that the wood might have nails hidden in it, but we found none. We did find metal flecks where the vat bands had deteriorated, but with wire brushes and large paint scrapers we removed almost all the rust. On our 16-in. radial-arm saw, we ripped the lumber to the rough sizes we needed, about ¼ in. thicker and ½ in. wider than the dimensions of the finished moldings. Next we prepared the stock on a jointer and a thickness planer. Once we had dressed down the old surfaces ¼ in., the wood was perfect and unmarked. As we worked,

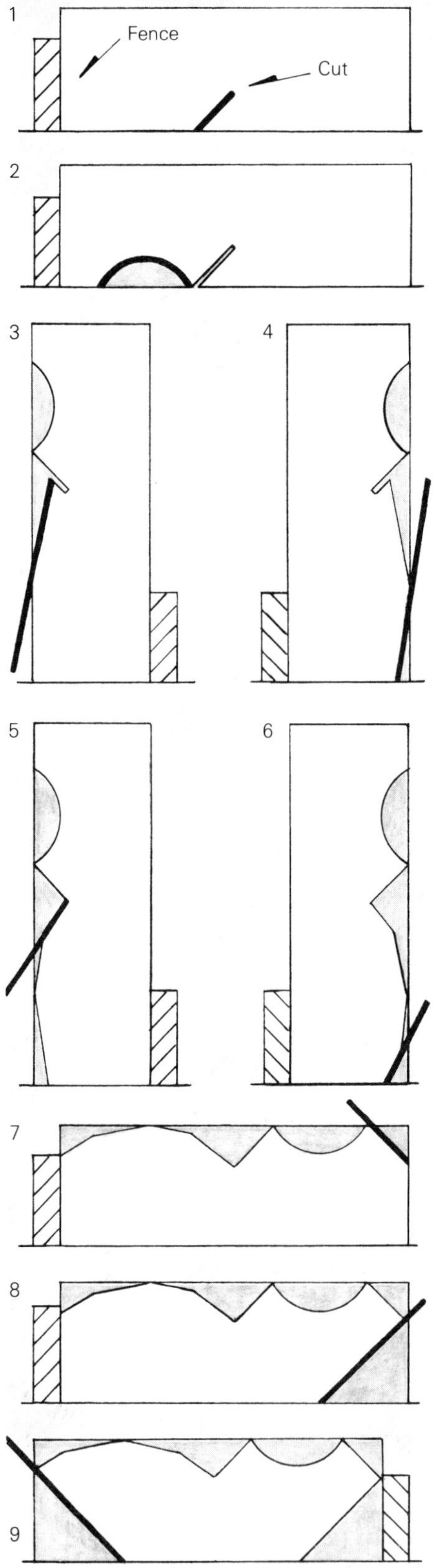

we checked our cutters for sharpness. Our stock was as straight and as square as we could make it; we were ready to begin shaping.

Setting up—Milling complex moldings on a table saw requires precision. Begin with an accurate template of the molding, to which you can adjust the sawblade's settings and against which you can compare results. The best template is a short piece of the molding you want to copy. If you must create a template from molding in place, you'll have to use a profile gauge. (See Figure 2 on the next page.) Many exterior moldings are too large to be handled with one application of the gauge. If this is the case with your trim, you'll have to take a series of readings, transfer them to paper and combine them for the complete profile. In fact, it's a good idea to sketch all molding profiles on site, for the gauges may get distorted before you return to the shop. Fashion your template out of a rigid material such as Masonite or plywood.

Before any cutting, even before setting the sawblade, scrutinize the template or molding cross section. The question is how to determine the order of the cuts. You don't want to take out a piece of stock you'll need later to run against the fence for making another cut. Think things through on a piece of paper. Certain cuts simply have to be made before others.

In Figure 1, for example, cut 1 is crucial because it is a dividing line between two curves: If its angle is incorrect or its cut misaligned, the proportions of both curves will suffer. If it is too deep, it undercuts the convex curve; if too shallow, material in the notch will have to be cleaned out later—a waste of time.

Cut 2, which creates the concave curve, must meet precisely the high point of cut 1. Because the stock is fed into the sawblade at an angle, this is a delicate cut.

Cuts 3 through 6, creating the convex curve, must be made after 2. If they had been made before 2, the convex curve would have made subsequent cuts a problem. (The stock could easily roll on that curve as it is fed into the sawblade.)

Cut 7 is delayed so that the point it creates with cut 2 won't be battered as the stock is maneuvered over the saw. Cuts 8 and 9 are made last, because leaving the corners of the stock square

Figure 1: Sequence of cuts

Illustrations: Carol Hubbard

From *Fine Homebuilding* magazine (April 1981) 2:48-49

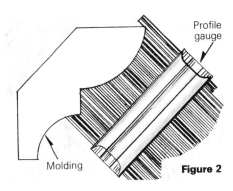

The Winooski Block (left) is capped by a cornice assembly over 6 ft. wide; it consists of 14 elements, including 7 moldings that were reproduced on a table saw. The milling of the molding is described on the facing page.

Profile gauge

Molding

Figure 2

You need a template to mill new moldings. Use a piece of the original, or transfer readings from profile gauge to paper on site and cut a template later in the shop.

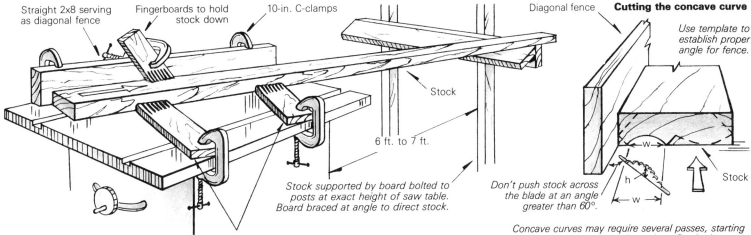

Straight 2x8 serving as diagonal fence

Fingerboards to hold stock down

10-in. C-clamps

Diagonal fence

Cutting the concave curve

Use template to establish proper angle for fence.

Stock

6 ft. to 7 ft.

Stock supported by board bolted to posts at exact height of saw table. Board braced at angle to direct stock.

Don't push stock across the blade at an angle greater than 60°.

Stock

Figure 3 Cutting setup

Fingerboards to hold stock against fence

Concave curves may require several passes, starting with the blade set low. On the last cut, saw points should just touch curve outline.

ensures the stability and accuracy of preceding cuts. (Cuts 3 through 6 would have been almost impossible if cut 9 had preceded them.)

To save time, pass all molding stock through a given saw setting; be fastidious about such settings, making practice cuts on scrap work. Cut more molding stock than you'll need at each setting, so you'll always have waste stock with the necessary previous cuts. In other words, to get an accurate setting for cut 5, you'll need stock with cuts 1 through 4 already made.

Cutting—We used a 10-in. Rockwell Unisaw with a 48-point carbide-tipped blade for all molding cuts. For most cuts we used the rip fence provided by the manufacturer. For cut 2 however, we needed a diagonal fence, so we trued a 2x8, used a template to carefully set it at the proper angle for the desired cove, and clamped it to the table with 10-in. Jorgensen C-clamps (Figure 3). To reduce stock flutter we used fingerboards, pieces of wood with a series of parallel kerfs cut in one end. Two fingerboards clamped to the table held the stock against the fence, while one fingerboard clamped to the fence held the stock down. The kerfs allowed enough play to let the wood slide through, but maintained enough pressure to ensure a straight cut. Using fingerboards and extension tables, you could cut all the molding unassisted, but you may prefer to have a helper to pull the stock gently through the last few inches of a cut. Several times a day, wipe the tabletop and sawblade clean with turpentine, to minimize binding.

Except for cut 2, all cuts were made with the

rip fence running parallel to the blade on one side or the other. As shown in Figure 1, cuts 1 and 2 were made with the stock face down on the table, while cuts 7, 8 and 9 were made with it face up. The stock stood on edge for cuts 3 through 5. (When cutting some symmetrical convex shapes, you can leave the sawblade at the same angle, and after one pass, turn the stock 180° to get the cut whose angle mirrors the first.) Each cut was preceded by carefully adjusting blade height and angle against the template. The last cuts on a molding (cuts 8 and 9) should be slightly larger than 45°—if your sawblade will tilt just a little more—to avoid gaps where the building surfaces are not quite perpendicular.

We cut the concave shape (cut 2) into the molding by passing the stock diagonally across the table-saw blade. (See Figure 3, at right.)To set the blade and fence correctly, you'll need a piece of the old molding. (A template is less effective.) Holding the high point of the curve over the blade, slowly crank up the blade so that the tip of the highest tooth just grazes that curve's apex; lock the setting and try a few cuts. To create the width of the curve, angle the piece of molding until all the teeth of the exposed blade lightly touch the arc of the molding. Another person should snug the fence against the angled molding and then clamp the fence to the table while you hold the molding in place. You'll have to tinker a bit to get the exact angle you need.

You can create almost any symmetrical curve with this method. Pushing the stock across the blade at a wider angle will result in a wider curve. However, the widest angle at which we

would push wood across the blade is 60°; with wider angles, not enough of the sawteeth are gripping and the blade will bind. (I'm not sure why, but a 48-tooth carbide blade binds up less than an 82-tooth one. It may be that the chips clear more easily.) If the blade is binding, make several passes to get the curve, starting the blade low and cranking it up ¼ in. for each pass. Don't get so wrapped up in your calculations that you become careless. Keep fingers clear of the blade. The speed at which you feed the stock must be determined on the job: Too fast and the blade will bind, too slowly and the wood will burn. The greater the angle of feed, the more often you should clean the blade.

The quality of the wood greatly affects the complexity of cuts you can make. Hardwoods are more difficult to mill without proper equipment. If concave curves are possible at all on hardwood, you'll have to make many gradually increasing cuts; the angle of the stock to the blade will be limited. Fortunately the grain in our cypress varied less than ¼ in. in 15-ft.

To refine the shape of our convex curves, we used many tools, including jack planes, curved shavehooks and spokeshaves. Among power sanders, Rockwell's Speed-block was the favorite; we clamped the finished molding to benches and sanded it using 50-grit pads.

Using these techniques we milled 1,000 linear feet for each of nine molding types, some more complex than the one described above. □

Bruce Andrews is a partner of Moose Creek Restorations Ltd., in Burlington, Vt.

Architectural Ceramics
Custom detailing from earth, fire and imagination

by Peter King

The crafts of custom woodworking, stained glass and stonemasonry have all enjoyed a revitalization during the last decade, and have all had a creative impact on residential construction. But ceramics has lagged behind. The great tilemakers of the American Arts and Crafts movement, men like John G. Low, Karl Langenbeck, Herman Mueller, Henry Mercer and William H. Grueby, pioneered work in the field of architectural ceramics, but their enterprises were short-lived. They were too few to leave behind an enduring legacy of their craft, and as a result custom tilemakers today are having to create an appreciation for their work, as well as having to rediscover or invent anew methods and processes for hand-producing high-quality ceramics for home building.

Most contemporary potters find that it is a lot easier to make a living producing functional stoneware, but for me and my partner Kathy Allen, the challenge of doing work on an architectural scale is truly rewarding. And we have discovered during the last several years that other ceramists around the country are beginning to take up that challenge as well.

The possibilities for using custom ceramics in a home are unlimited: decorative doorways, fancy fireplaces, outrageous bathrooms. Used with restraint as an appropriate accent, architectural ceramic work can be very pleasing to the eye, very durable as a finish surface and, as custom work goes, relatively economical.

Houses and clients—Custom ceramics can be used to advantage in those areas of a home that are traditionally decorated with mass-produced tile, with commercially available porcelain products, plastic laminate or stone. These include bathroom sinks, vanities, bathtubs and floors, kitchen counters and backsplashes, along with fireplace fronts and hearths. But we have also explored the use of ceramics in other places, some of them unorthodox, including foyers and wall fountains, patio murals, door casings, woodstoves and terrace walls.

Most of our clients are home owners rather than builders or architects. In northwest Florida, tastes run to the conservative side, which means that most builders and architects are reluctant to incorporate unusual detailing into the spec and custom houses they build. As a result, it is the

Peter King and Kathy Allen are partners in Stonehaus, an architectural-ceramics company and studio in Pensacola, Fla.

owners who first contact us, and who urge the builder or architect to help find a place for our work in their homes. Thus the design proposals we make must take into account the final interior appearance of the home, down to the furniture and other detailing.

We begin our design work by looking at the working drawings. An exterior elevation can do a lot to steer even an interior fireplace design in the right direction. We then discuss the client's expectations for the rest of the interior. Although each job we do is unique, we now have a fairly extensive portfolio, and it helps to leaf through it with our clients to determine their likes and dislikes with regard to our own work. We often ask them to pull pictures of things that appeal to them from books and magazines, even if these have nothing to do with ceramics. We have executed pieces inspired by jukeboxes, woodwork, carpets, fabrics and nature prints. The design for the fireplace front shown in this article grew out of the client's liking for Southwest Indian art, and for birds.

Once we have a clear idea of what is appropriate to the job, we draw up several different design proposals. These may be variations on a given theme or radically different treatments of the same idea.

Meeting again with our clients, we review the options and, usually at this time, decide on a

given design. The clients may suggest changes, which we incorporate. If possible, we also have our clients view the work and discuss it with us after their piece is made but while the clay is still wet, as this is the last possible time at which change can be made.

Using this method of discussion and review, our clients have always been satisfied, even if they did not have a clear vision of the final product at the beginning of the project. With plans clearly made and designs approved at every stage of the work, no one is shocked by the results. This is very important because, unlike wood and metal, ceramic material cannot be altered. It's an all-or-nothing proposition once the clay is in the fire.

Because the architectural ceramists are the "aliens," the new guys on the job site, it is extremely important for them to establish a good rapport with the other subcontractors. We are fortunate to have worked extensively in the building trades before becoming potters. But ceramists who have no experience in residential construction and want to do architectural work will be all right as long as they rely on the expertise of the tilesetters and masons already on the job. At the very least, a potter should be able to read building plans and discuss general construction techniques with all others involved in the project. A good relationship with the other subs goes a long way in ironing out the wrinkles of any custom job.

Tools of the craft—The rapid growth of studio pottery in the last 15 years has spawned an entire industry of pottery-equipment manufacturing. As a result, potters now have access to a wide range of very sophisticated equipment. While some of these tools can be found in almost any pottery studio, there are others which, until recently, have not been widely available.

For example, the potter's wheel has always been a fixture in every ceramics studio, but not the slab roller. This is an ingeniously simple device that looks like the wringers on an old-time washing machine and is used to make large sheets of clay that are of uniform compression and thickness. While many potters use the slab roller to simplify the construction of pieces that could be done by more primitive methods, we use this tool to create ceramic art with architectural applications.

Before the studio version of this machine was introduced, potters made slabs by pounding and hand-rolling the clay. But in large pieces, clay

From *Fine Homebuilding* magazine (August 1987) 41:70-75

flattened by hand cannot be uniformly compressed, and its inconsistent density can result in tiles that warp and crack when they are fired. The slab roller eliminates this problem, and also produces clay sheets with the uniform thickness of milled lumber, on a scale not possible with hand processes. This simple device thus enables the studio potter to turn out large-scale work with a level of refinement that was previously nearly impossible.

Small pug mills with extrusion dies enable contemporary potters to make custom ceramic moldings at the studio level. Every studio also needs a clay mixer. The large vertical-shaft mill we use (photo top left), can blend fairly stiff clay in short order. I had it made up from cast-off industrial components because nothing so sturdy and so powerful was commercially available for studio potters.

Preparing the clay—The actual work involved in making architectural ceramics is similar to that for making smaller ceramic pieces, but on a larger scale. Vast amounts of clay have to be shaped, textured and glazed, and weighty pieces have to be fired and installed. But many of these problems can be overcome by a careful choice of materials.

Though there are new exotic clay bodies on the market, some of which have architectural applications, we use a more traditional approach. As our home area of northwest Florida has many excellent clay deposits, we choose to mine our own raw materials. The clay we mine is an excellent kaolinic stoneware clay. Kaolin gets its name from a hill in China that yielded up a fine white clay. In the making of pottery, we use this excellent material just as it comes out of the ground.

But to adjust this clay for architectural work, we add 25% to 30% grog—a material that is basically ground-up firebrick. This does several things to render the clay more suitable to our purpose. All clay has a wet-to-dry shrinkage plus an unfired-to-fired shrinkage, and the addition of a pre-fired, vitreous material reduces both these rates of shrinkage. From the standpoint of design and layout, we can shape ceramic components from green clay that are fairly close to their finished size and proportion (the shrinkage rate is about 10%). From a structural point of view, less shrinkage means the pieces are less likely to warp and crack in the drying and the firing stages. A similar clay body can be made by the addition of 25% to 30% grog to any good stoneware clay.

To blend large batches of clay, Peter King designed and built a vertical-shaft pug mill (above left). With its 5-hp motor, heavy-duty transmission and specially designed paddle, it works like a giant mixer to turn out enough clay for an architectural project. To make a large ceramic panel, 1-in. thick sheets of clay are produced by a simple device called a slab roller. It looks something like a washing-machine wringer, and ensures that the clay is of a uniform consistency. The clay slabs (left) are then placed on a concrete floor that has been sprinkled with grog (a powder made from ground firebrick). Then it is joined with other slabs to make up the whole rough panel.

Incising the pattern. Once the clay panel has been dimensioned, the design is traced onto it by cutting through the pattern with a knife (left). These knife lines are enlarged and shaped with whatever tool is appropriate—here the tip of a brush handle (below). Finally, the panel is cut into individual tiles (below left).

Shaping and firing—Most of our architectural work begins with the making of slabs. For a fireplace front or mural we make 1-in. thick slabs of grogged clay about 30 in. long by 40 in. wide (bottom photo, facing page). The slabs are lapped together on a concrete floor that has been sprinkled with a coating of grog. The grog allows the clay panel to slide on the floor as it shrinks during the drying stage. The overlaps of the panels are held to about 2 in., and the entire composite panel is then hand-pounded and rolled with a large rolling pin.

This leaves us with a single clay sheet of uniform thickness. We have made panels up to 30 ft. long using this technique. By laying the slabs over specially shaped pieces of wood or over plaster forms set on the floor, we can provide high relief or contours to the panel, and still retain a continuity of appearance. Due to the clay's shrinkage we must add the shrinkage rate

to the overall dimensions of a custom-fitted piece. Once the rough panel is complete, we cut it to shape and overall dimensions. For shallow relieved areas on a given design, we can then add more clay to the slab or, in some cases, carve clay away.

On the project shown here, we incised the design into the clay by laying the paper pattern over it and cutting right through with an X-acto knife (photo top left). These thin lines were enlarged with a stylus or other tool of the correct shape for the effect we wanted (photo above right). Usually the tip of a brush handle works just fine. When the design work is completed, we cut the entire panel into sections or individual tiles that can be handled through the remainder of the pottery process. These sections are cut freehand in the wet clay with an eye to design (the grout joints of a finished ceramic panel are often an important part of the

design), as well as ease of handling. The panel is then allowed to dry on the floor. Complete drying takes two to four weeks, depending on the weather conditions. This can be a problem when working with tight construction schedules.

When the panel is completely dry, all the edges of each piece are hand-sanded to remove any burrs or slight irregularities. We then reassemble it and glaze it as a single piece to ensure a uniform overall finish.

Next comes firing. We fire most of our work in a 60-cu. ft. kiln that's fueled by natural gas and made of hard firebrick. The greenware is loaded into the kiln much as baked goods are shelved in a large oven. The pieces cannot be stacked or even touch one another or they will fuse together when the glaze melts. Each piece is placed separately on a silicon-carbide shelf. These shelves can bear up without deflecting or collapsing under extremely high heat. We spread

Installing the panel. To prepare drywall for the cement bed, expanded-metal lath is stapled securely to the studs (top). After the thinset wall mix is troweled onto the metal lath, the tiles are pressed into the bed, beginning at the bottom (above). After mushing several tiles into the thinset, King and his partner Kathy Allen install shims between the pieces to create even grout lines (right).

grog under each piece so that it can contract during firing without sticking to the shelf. When the kiln is fully loaded, we seal the doorway with a stack of firebricks.

Because we make such large, thick pieces, our firings take longer than those for ordinary ceramic ware. If a tile is heated too rapidly, its outer surface will begin to sinter and harden, while the interior still contains chemically bound water. When the interior of the piece heats up, this bound water produces a lot of

steam and pressure, and the tile explodes. It sounds like someone slamming a car door inside the kiln.

A stoneware potter can afford to lose a piece this way every now and then, but we can't. For architectural work each piece is an integral, irreplaceable part of the whole. If one blows up, the whole job is ruined. This is why we allow for plenty of time during the first 1,000°F of temperature increase. It usually takes 24 to 36 hours for the kiln to reach the desired temperature—

usually 2,400°F. Once this temperature is achieved, the damper is closed and fuel is shut off. We wait at least 24 hours before opening the kiln.

To achieve certain special effects or specific glaze colors, we have to fire the clay at lower temperatures, and for this we often use our three electric kilns. Bright colors are more easily produced at lower temperatures, and colors can be more easily controlled in the oxidizing atmosphere that electric kilns provide. Kilns fired by

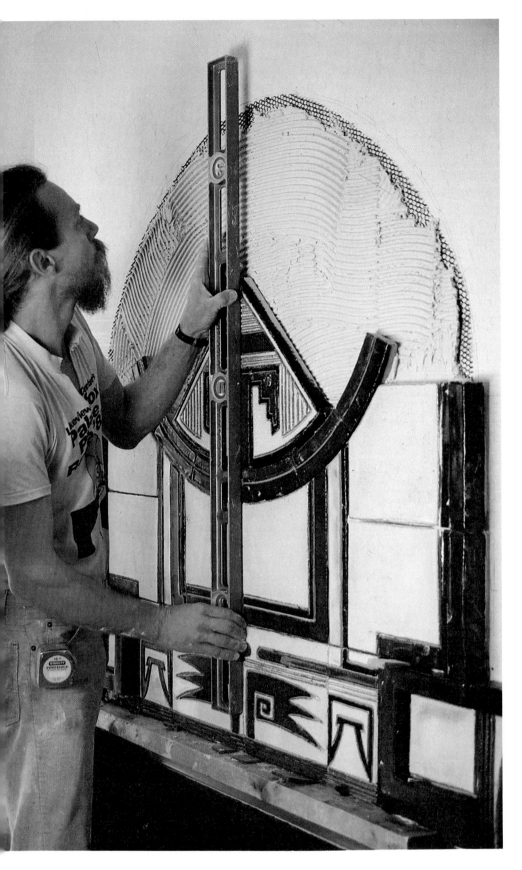

The tiles that form the bird motif have to be plumbed (left). After the joints are grouted, the surface of the tile is wiped down first with a vinegar solution, then with water, to remove the residual haze left behind by grouting compound (above).

combustion consume most of the oxygen, and leave little free to combine with the glazes.

If insufficient oxygen is present in the kiln to allow all of the fuel to burn completely, a reaction called reduction takes place. This happens when free carbon in the kiln atmosphere oxidizes by bonding with the available oxygen in the clay and the glazes. Reduction firing reliably produces a muted, stony-looking glaze. This may be desirable, depending on the potter's intent, or it may not.

Installation—Once a piece is finished, installation is done with fairly traditional tilesetting techniques. On most fireplace fronts we staple expanded-wire lath over gypsum wallboard (top left photo, facing page), taking care that the wire is firmly fastened to the stud frame. If the wall is brick, block or concrete, we can forego the wire and apply the thinset adhesive directly to the existing surface.

We then thoroughly bed the wire with thinset wall mix. Because of the thickness, size and in-

terlocking geometry of our pieces, we have found it unnecessary to do the customary preliminary coat of cement (the scratch coat) and let it dry in order to get a solid job. After the wall has been bedded, we apply additional thinset to the back of each ceramic component to ensure that no voids result from any irregularity of the mortar bed or the back of the tile.

We start setting the tile at the bottom of the panel (bottom left photo, facing page). We force each piece into the bed of thinset, slightly above its final position. Then we let each piece slide down onto the tile below it. After a number of tiles have been pressed into the bed, we insert shims between them to create uniform gaps for the grout lines (photo facing page, right). Allowing the pieces to slide up and down ensures complete contact between the tile and the wall, resulting in a strong, solid bond.

Some parts of a design, like the bird motif in the fireplace front shown here, have to be aligned and plumbed with a carpenter's level (photo left).

Once set, the panel is allowed to stand for a day before grouting. This is done with a sanded premixed grout of suitable color. On the third day, the finished panel is wiped down with 5% acid vinegar, then water, to remove any residual film left by the grouting (photo above).

The fireplace front, which started as a single slab of clay, is now reunified as a ceramic panel, with its various sections and grout lines part of its overall composition. □